Springer Proceedings in Materials

Volume 4

Springer Proceedings in Materials publishes the latest research in Materials Science and Engineering presented at high standard academic conferences and scientific meetings. It provides a platform for researchers, professionals and students to present their scientific findings and stay up-to-date with the development in Materials Science and Engineering. The scope is multidisciplinary and ranges from fundamental to applied research, including, but not limited to:

- Structural Materials
- Metallic Materials
- Magnetic, Optical and Electronic Materials
- Ceramics, Glass, Composites, Natural Materials
- Biomaterials
- Nanotechnology
- Characterization and Evaluation of Materials
- Energy Materials
- Materials Processing

To submit a proposal or request further information, please contact one of our Springer Publishing Editors according to your affiliation:

European countries: **Mayra Castro** (mayra.castro@springer.com)
India: **Akash Chakraborty** (akash.chakraborty@springernature.com)
Singapore, Southeast Asia and Oceania: **Ramesh Nath Premnath** (ramesh.premnath@springernature.com)
The Americas: **Michael Luby** (Michael.luby@springer.com)
China and all other countries or regions: **Mengchu Huang** (mengchu.huang@springer.com)

More information about this series at http://www.springer.com/series/16157

Raul Fangueiro · Sohel Rana
Editors

Advanced Materials for Defense

Development, Analysis and Applications

Springer

Editors
Raul Fangueiro
Department of Mechanical Engineering
School of Engineering
University of Minho
Guimarães, Portugal

Sohel Rana
University of Huddersfield
Huddersfield, UK

ISSN 2662-3161 ISSN 2662-317X (electronic)
Springer Proceedings in Materials
ISBN 978-3-030-34122-0 ISBN 978-3-030-34123-7 (eBook)
https://doi.org/10.1007/978-3-030-34123-7

This Springer imprint is published by the registered company Springer Nature Switzerland AG
The registered company address is: Gewerbestrasse 11, 6330 Cham, Switzerland

Scientific Committee Members (AuxDefense 2018)

List of Participants (AuxDefense 2018)

Adam Foltz, USA, email: adam.l.foltz.civ@mail.mil
Adam Nevin, UK, email: adam.nevin@trelleborg.com
Adam Puszkarz, Poland, email: adam.puszkarz@p.lodz.pl
Adrianna Mackiewicz, Poland, email: addrianna10@gmail.com
Alexis Vilaça, Portugal, email: alexisvilaca@pixartidea.com
Aloysio Gomes de Souza Filho, Brazil, email: asouzafilho@gmail.com
Amândio Besteiro, Portugal, email: arbesteiro@hotmail.com
Amaya Igartua, Spain, email: amaya.igartua@tekniker.es
Ana Fangueiro, Portugal, email: ana.fangueiro@sciencentris.com
Ana Ferreira Azevedo, Belgium, email: ana.ferreira@dymasec.be
Ana Leitão, Portugal, email: analeitao@fibrenamics.com
Ana Maria Botelho do Rego, Portugal—Instituto Superior Técnico, email: amrego@tecnico.ulisboa.pt
Ana Maria Ferraria, Portugal—Instituto Superior Técnico, email: ana.ferraria@tecnico.ulisboa.pt
Ana Maria Rocha, Portugal
Ana Roças, Portugal, email: ana.rocas@lma.pt
André Ferreira, Portugal, email: andrefilipegoncalvesferreira@gmail.com
André Pepe, Portugal, email: andrepepe@fibrenamics.com
Andrew Johnson, UK, email: a.johnson@lboro.ac.uk
Anna Mina, Cyprus, email: mina.anna@ucy.ac.cy
António Pereira Gonçalves, Portugal, email: apg@ctn.tecnico.ulisboa.pt
Arlindo Santos, Portugal, email: norberto.almeida@fibrauto.pt
Arthur Voisin, France, email: arthur.voisin@volvo.com
Arvind Kumar Bhakta, Belgium, email: arvind.bhakta@unamur.be
Augusto Brito, Portugal, email: ramos.brito@marinha.pt
Balakrishnan Marimuthu, Portugal, email: balki2009@yahoo.com
Beatriz Ferreira, Portugal, email: fnunes@2c2t.uminho.pt
Behjat Ansari, UK, email: behjat.ansari@bristol.ac.uk
Bruno Marinho, Brazil, email: bcmarinho@gmail.com
Bruno Serrano, Portugal, email: baserrano@emfa.pt

Burigede Liu, UK, email: BL377@cam.ac.uk
Carina Cunha, Portugal, email: carinacunha@fibrenamics.com
Carlos Mota, Portugal, email: cmota@tecminho.uminho.pt
Catarina Pimenta, Portugal—UMinho, email: catarina.rpimenta@hotmail.com
Chee Lip Gan, Singapore, email: clgan@ntu.edu.sg
Clara Gonçalves, Portugal, email: claracpgoncalve@gmail.com
Claudia Costa, Portugal, email: claudia.costa@surforma.com
Claudiu Pavel, The Netherlands, email: claudiu.pavel@ec.europa.eu
Clementina Freitas, Portugal, email: cf@latinogroup.net
Daiva Milasiene, Lithuania, email: daiva.milasiene@ktu.lt
Damjan Cekerevac, Portugal, email: damjan.cekerevac@uc.pt
Danica Simic, Sérvia, email: simic_danica@yahoo.com
Daniel Ambrosini, Argentina, email: dambrosini@uncu.edu.ar
Danielle Beduschi, Portugal—UMinho, email: beduschi.dani@gmail.com
David Lange, Portugal, email: davidlange@pixartidea.com
Denis Josse, France, email: denis.josse@sdis06.fr
Diana Ferreira, Portugal, email: diana.ferreira@det.uminho.pt
Diana Oliveira, Portugal, email: dianaoliveira@fibrenamics.com
Edite Felgueiras, Portugal, email: editefelgueiras@fibrenamics.com
Enio Chambel, Portugal—Academia Militar, email: enio.chambel@academiamilitar.pt
Fausto Mendonca, Brazil, email: faustobm@hotmail.com
Fernanda Steffens, Brazil, email: fernanda.steffens@ufsc.br
Fernando Cunha, Portugal, email: fernandocunha@det.uminho.pt
Fernando Ferreira, Portugal, email: fnunes@2c2t.uminho.pt
Fernando Ribeiro Oliveira, Brazil, email: oliveira.fernando@ufsc.br
Filipa Fernandes, Portugal, email: fdfernandes@oozenanotech.com
Filipe J. Oliveira, Portugal—University of Aveiro, email: filipe@ua.pt
Filipe Nuno Moreira, Portugal, email: filipemoreira@fibrenamics.com
Filipe Soutinho, Portugal, email: filipe.soutinho@idtconsulting.pt
Francisca Martínez, UK, email: fmartin2@exseed.ed.ac.uk
Francisco Galindo-Rosales, Portugal—FEUP, email: galindo@fe.up.pt
Francois Boussu, France, email: francois.boussu@ensait.fr
Gilda Santos, Portugal—CITEVE, email: gsantos@citeve.pt
Giuseppe G. Daquino, Belgium, email: giuseppe.daquino@eda.europa.eu
Hugo B. Rebelo, Portugal, email: h.rebelo@campus.fct.unl.pt
Inese Ziemele, Latvia, email: inese.ziemele@rtu.lv
Ionelia Voiculescu, Romania, email: ioneliav@yahoo.co.uk
Isabel Duarte, Portugal—University of Aveiro, email: isabel.duarte@ua.pt
Israel Gago, Spain, email: israel.gago@upct.es
Ivan Medeiros, Brazil, email: ivanmedeiros@ct.ufrn.br
Ivo Almeida, Portugal, email: ivoazorean@gmail.com
Ivo Cruz, Portugal, email: ivocruz@fibrenamics.com
Jakub Ederer, Czech Republic, email: jakub.ederer@ujep.cz
Jan Andzelm, USA, email: jan.w.andzelm.civ@mail.mil

Jindrich Fornusek, Czech Republic, email: jindrich.fornusek@fsv.cvut.cz
Jiri Henych, Czech Republic, email: henych@iic.cas.cz
Joana Rompante, Portugal, email: joanarompante@fibrenamics.com
João André Pereira, Portugal, email: joaopereira@fibrenamics.com
João Bessa, Portugal, email: joaobessa@fibrenamics.com
João Caetano, Portugal, email: jvcaetano@academiafa.edu.pt
João Filipe Pereira, Portugal, email: joaofilipe@fibrenamics.com
João Madeira, Portugal, email: joao.madeira@defesa.pt
João Motta, Portugal, email: graca.mota@marinha.pt
João Piedade, Portugal, email: lourenco.piedade@marinha.pt
Joaquim M. Vieira, Portugal—University of Aveiro, email: jvieira@ua.pt
Jörg Stierstorfer, Germany, email: jstch@cup.uni-muenchen.de
Jorge Ribeiro, Portugal, email: jorgeribeiro@fibrenamics.com; ribeiro.jmg@mail.exercito.pt
Jos Brouwers, The Netherlands, email: jos.brouwers@tue.nl
José Antonio Loya, Spain, email: jloya@ing.uc3m.es
José Borges, Portugal, email: jose.borges@academiamilitar.pt
José Covas, Portugal—UMinho, email: jcovas@dep.uminho.pt
José Gisbert, Spain, email: jgisbert@aitex.es
Julian Carretero, Spain, email: jcnciney@gmail.com
Júlio Sérgio Dolce da Silva, Brazil, email: jsdolce@gmail.com
Jun Lou, USA, email: jlou@rice.edu
Lars Montelius, Portugal, email: cristina.louro@inl.int
Leandro Lopes, Portugal, email: leandrolopes@fibrenamics.com
Liliana Martelo, Portugal, email: liliana.martelo@tecnico.ulisboa.pt
Lionel Gilson, Belgium, email: lionel.gilson@mil.be
Lisandra Vilaça, Portugal, email: lisandra.vilaca@smartinovation.com
Lorenzo Moroni, Italy, email: lorenzo.moroni@leonardocompany.com
Luca Dely, Hungary, email: delyluca@gmail.com
Luís Arsénio, Portugal, email: luis.arsenio@gmail.com
Luis Félix, Portugal, email: lffelix@academiafa.edu.pt
Luís Filipe Pratas Quinto, Portugal, email: luis.quinto@academiamilitar.pt
Luis Madson Silva Costa, Brazil, email: madsonfisica@gmail.com
Luís Nobre, Portugal, email: luisnobre@fibrenamics.com
Luís Pereira, Portugal, email: lfpereira@emfa.pt
Luís Pina, Portugal—INEGI, email: lpina@inegi.up.pt
Maciej Panek, Poland, email: maciej.panek@pwr.edu.pl
Manuel Barros, Portugal, email: manuel.barros@lma.pt
Manuel Cruz, Portugal
Manuel del Río, Spain, email: manurigo96@gmail.com
Marcin Landwijt, Poland, email: mlandwijt@moratex.eu
Marco Pinto, Portugal, email: mfmpinto@gmail.com
Maria Conceição Paiva, Portugal—UMinho, email: mcpaiva@dep.uminho.pt
Maria José Ferreira, Portugal, email: fnunes@2c2t.uminho.pt

Mario Berberan Santos, Portugal—Instituto Superior Técnico, email: berberan@tecnico.ulisboa.pt
Mario Emanuele de Simone, UK, email: m.e.de.simone@bath.ac.uk
Marta Brito, Portugal, email: martabrito@fibrenamics.com
Matej Vesenjak, Slovenia, email: matej.vesenjak@um.si
Michal Mara, Czech Republic, email: michal.mara@fsv.cvut.cz
Michele Buonsanti, Italy, email: michele.buonsanti@unirc.it
Michele Meo, UK, email: m.meo@bath.ac.uk
Nandan Khokar, Sweden, email: nandan.khokar@biteam.com
Nejc Novak, Slovenia, email: n.novak@um.si
Nelson Matos, Portugal, email: ncmatos@isq.pt
Nicolis Araujo, Brazil, email: nicolis.amaral@yahoo.com.br
Norberto Almeida, Portugal, email: norberto.almeida@fibrauto.pt
Nunzio Cennamo, Italy, email: nunzio.cennamo@unicampania.it
Panagiotis Kavouras, Greece, email: kavouras@chemeng.ntua.gr
Paul Kiekens, Belgium, email: match@ugent.be
Paulo Antunes, Portugal, email: pjantunes@critical-materials.com
Paulo Reis, Portugal—University of Beira Interior, email: preis@ubi.pt
Paulo Simões, Portugal, email: paulosimoes@pixartidea.com
Pavel Janoš, Czech Republic, email: pavel.janos@ujep.cz
Pedro Neto, Portugal, email: neto.pmc@mail.exercito.pt
Pedro Semide, Portugal, email: teodoro.semide@marinha.pt
Pedro Silva, Portugal, email: pedrosilva@fibrenamics.com
Pedro Tomás Silva, Portugal, email: silva.pmt@mail.exercito.pt
Petra Pötschke, Germany, email: poe@ipfdd.de
Ping Song, China, email: sp2016@njust.edu.cn
Rafael Basto, Portugal, email: rpbasto@emfa.pt
Raul Fangueiro, Portugal, email: rfangueiro@dem.uminho.pt
Rebecca Mangham, UK, email: rrmangham@dstl.gov.uk
Remigijus Ivanauskas, Lithuania, email: remigijus.ivanauskas@ktu.lt
Ricardo Andrade, Brazil, email: ricardo.andrade@mackenzie.br
Ricardo Pinto, Portugal—INEGI, email: rpinto@inegi.up.pt
Robert Chapple, UK, email: rc10aes@bolton.ac.uk
Robert Young, UK, email: robert.young@manchester.ac.uk
Rodrigo Lourenço, Portugal, email: ralourenco@surforma.com
Rui Costa, Portugal, email: rucosta@fc.up.pt
Rui Magalhães, Portugal, email: ruimagalhaes@fibrenamics.com
Samuel Migueis, Portugal, email: migueis.sc@mail.exercito.pt
Sara Silva, Portugal, email: sarasilva@det.uminho.pt
Seif Sadek, Portugal—INEGI, email: ssadek@inegi.up.pt
Sérgio Branco, Portugal, email: asergio.branco@gmail.com
Shama Parveen, UK, email: Parveenshama2011@gmail.com
Sheila Young, UK, email: robertandsheila@btinternet.com
Shijie Wang, Singapore, email: sj-wang@imre.a-star.edu.sg
Shripad Mahulikar, India, email: spm@aero.iitb.ac.in

Shuai Chen, China, email: shuai_chen@aliyun.com
Sofia Costa, Portugal, email: sofiacosta@fibrenamics.com
Sohel Rana, UK, email: s.rana@hud.ac.uk
Sónia Costa, Portugal, email: soniacosta@fibrenamics.com
Stefan Drawin, France, email: stefan.drawin@onera.fr
Thiago Felix dos Santos, Brazil, email: thiagotextilufrn@hotmail.com
Tiago Pereira de Sousa, Portugal, email: tiagosousa@fibrenamics.com
Torsten Laessig, Germany, email: andrea.seidel@emi.fraunhofer.de
Uli Heisserer, The Netherlands, email: ulrich.heisserer@dsm.com
Vasco Prates, Portugal, email: marques.prates@marinha.pt
Vera Gonçalves, Portugal, email: verareisgoncalves@gmail.com
Wiesława Urbaniak-Domagała, Poland, email: wurbando@p.lodz.pl
Wilson Antunes, Portugal, email: antunez.wdta@gmail.com
Y. Y. Y. Cao, The Netherlands, email: y.cao.1@tue.nl
Yousef Dobah, UK, email: y.dobah@bristol.ac.uk
Yuri Merkulov, Russia, email: melas33@yandex.ru
Zsolt Végvári, Hungary, email: kraius@kraius.hu
Jaap Weerheijm, The Netherlands, email: jaap.weerheijm@tno.nl
Buonomo Monica, Itália, email: nunzio.cennamo@unicampania.it
Cennamo Maria Francesca, Itália, email: nunzio.cennamo@unicampania.it
Mário Martinho, Portugal
Mikhail Mokhov, Portugal, email: gabdefembassy@sapo.pt

Preface

This book is a collection of selected high-quality research papers submitted to *AuxDefense 2018—1st World Conference on Advanced Materials for Defense* held in Lisbon, Portugal, covering various topics related to defense applications.

AuxDefense 2018 was the meeting point for everyone interested in these outstanding advanced materials, including researchers, technicians, technologists, militaries and students. This conference focused on advanced materials research in the forward-looking R&D domains where the intersection of key technologies in areas such as nano and micro-technology, biotechnology, stealth materials, smart materials and structures and energy generation and storage is being explored.

This book is focused on advanced materials for defense applications and mainly covers *Raw Materials, Advanced Polymers, Textiles and Composites, Experimental Characterization, Modeling* and a few *Advanced Applications*. This book discusses about different raw materials used in the European defense industries, molecularly imprinted polymers for security applications, advanced 3D and auxetic textile structures for impact and ballistic protection, advanced high performance composites reinforced with carbon nanotubes and other nanomaterials developed for defense applications, high performance cementitious composites for defense infrastructures, modeling of different behaviors such as impact, ballistic and blast using advanced simulation methods and lastly, important issues related to a few important defense applications including personal protection, blast protection, small caliber weapons, detection of chemical vapors, etc.

This book provides an extensive collection of latest research and development taking place in the area of defense by the scientists and academics from multi-disciplinary areas including polymers, textiles, chemistry, mechanical engineering, engineering mechanics, nanotechnology, civil engineering, etc. *The R&D works covered in this book aim to improve the safety and security as well as comfort of the people working in the defense sector.* Therefore, this book will not only benefit the scientists and researchers to gather new knowledge but also strongly benefit the engineers from different industries involved in the production of defense materials as well as people in the defense industries who are the targeted users of these novel technologies. Therefore, we strongly recommend this book to the Master and Ph.D. students, researchers, academics, scientists, engineers and R&D staff of defense-related industries.

Guimarães, Portugal Raul Fangueiro
Huddersfield, UK Sohel Rana

Contents

About the Editors

Prof. Raul Fangueiro is currently a professor in the Department of Mechanical Engineering and senior researcher in the Center for Textile Science and Technology, at the University of Minho, Portugal. He is the head of the Fibrous Materials Research Group of the same university with expertise in advanced materials (nano, smart, composites) and structures (3D, auxetic, multiscale) with 45 researchers. He is the mentor and the coordinator of the **FIBRENAMICS International Platform** including 300 partners developing promotion, dissemination, technology transfer and research activities on fiber-based advanced materials. He has more than 160 published papers in international reputed scientific journals, 420 conference publications, 36 books and 20 patents. He coordinated more than 50 national and international research projects on advanced fibrous and composite materials, mainly for defense, building, architectural and health-care applications. He supervised 20 Ph.D. and Post-doc scientific works and is a member of the editorial board of several leading international scientific journals on composite and fibrous materials. Raul Fangueiro is the scientific coordinator of *ICNF—International Conference of Natural Fibers* and *AuxDefense—Advanced Materials for Defense* and is an active member of the working groups of the European Defense Agency and NATO.

Dr. Sohel Rana is currently a lecturer in Technical Textiles in the School of Applied Sciences, University of Huddersfield, UK. Previously, he worked as Senior Researcher and Leader of Composite Materials at the Fibrous Materials Research Group (Fibrenamics), University of Minho, Portugal (2010–2018). He obtained his bachelor's degree in Textile Technology from the University of Calcutta, India (2002), and master's degree (2005) and Ph.D. in Fiber Science and Technology (2011) from Indian Institute of Technology (IIT, Delhi), India. Dr. Rana is actively involved in the research and development of smart fibrous and composite materials, nanocomposites, bio-composites, natural fibers, electrospinning, etc. He has been a key researcher in developing 'multi-functional and biobased hierarchical composites' and high performance 'auxetic structures and composites' for personal protection applications. He has been a mentor for several Ph.D. and post-doc students at the University of Minho. Benefited from his long experience, Dr. Rana has edited a number of books on advanced textile and composite materials with the leading publishers, among which *Advanced Composites in Aerospace Engineering* (Woodhead Publishing), *Braided Structures and Composites* (CRC Press), *Fibrous and Textile Materials for Composite Applications* (Springer) received well recognition within the scientific community. To his credit, Dr. Rana has published over 35 papers in peer-reviewed international journals and contributed to 14 book chapters and more than 60 papers in international conferences.

Dr. Rana is a former member of the European Society for Composite Materials and served as a member of the organizing and scientific committee of many international conferences (e.g., ICNF 2015, ICNF 2017, ICNF 2019, etc.). He has been invited to deliver keynote and invited talks in several international conferences and meetings focused on fiber science and composite materials. He also participates in the guest editorial board of several scientific journals (e.g., Journal of Nanomaterials, Key Engineering Materials, International Journal of Polymer Science, Journal of Biobased Materials and Bioenergy, Advances in Polymer Technology, etc.) and is a potential reviewer for numerous scientific journals including Green Chemistry, Carbohydrate Polymers, Composite Science and Technology, Composites Part A & B, Dyes and Pigments, RSC Advances, Composite Interfaces, Journal of Composite Materials, Journal of Reinforced Plastics and Composites, Powder Technology, Journal of Nanomaterials, Journal of Applied Polymer Science, Fibres and Polymers and so on.

Raw and Processed Materials Used in the European Defence Industry

Claudiu C. Pavel, Evangelos Tzimas, Patrícia Alves Dias and Darina T. Blagoeva

Abstract In the context of the EU raw materials strategy, this work identifies the raw and processed materials important to the European defence industry. 47 processed materials, including different alloys, composites and compounds were identified as highly important for the manufacturing of defence applications systems. A further assessment based on the composition of these 47 high-performance processed materials indicates that 39 raw materials are necessary for their production. Out of these 39 raw materials, 20 are currently critical for the EU economy and have associated potential supply risk.

Keywords Defense applications · Raw materials · Alloys · Composites

1 Introduction

As the technologies become more sophisticated, the social expectation for better performance continues to grow. All of this is underpinned by many raw and advanced materials that provide unique properties to products. New materials, whether they are based on rare earth elements (REEs) or lightweight composites, enable the modern technologies to be smaller, lighter and more energy efficient. A significant part of the manufacturing industries and service economies are now, in turn, becoming more dependent on these materials.

Europe imports many raw materials to meet its basic needs and therefore relies on a global matrix of various materials supply chains. This reliance is growing and with it one of the EU's biggest challenges in the years to come. At the same time, very little primary production occurs within the EU's member states.

To ensure a reliable supply of raw materials, the European Commission is investigating which of them are particularly important for the EU economy and assesses their supply risk. This has enabled to identify a set of so-called *critical raw materials* (CRMs). Three assessments were performed by now and the latest study published

C. C. Pavel (✉) · E. Tzimas · P. A. Dias · D. T. Blagoeva
European Commission, Joint Research Centre, Energy, Transport and Climate Directorate, Westerduinweg 3, 1755 LE Petten, The Netherlands
e-mail: claudiu.pavel@ec.europa.eu

R. Fangueiro and S. Rana (eds.), *Advanced Materials for Defense*, 4, Springer Proceedings in Materials, https://doi.org/10.1007/978-3-030-34123-7_1

in 2017 found that 27 raw materials are deemed critical for civil applications at the EU level [2].

Although the critical raw materials are in general not used in massive quantities, they are essential for applications such as in energy and defence sectors. The situation becomes even more acute with regard to imports of specialty materials used in high-technology applications related to the defence sector where security and operational autonomy are essential.

As many other economic sectors, the European defence industry relies on the use of a wide range of materials with unique properties that make them essential for the manufacture of components used in military applications because the use of substitutes does not always guarantee the same performance. For example, REEs are indispensable in remotely piloted aircraft systems, precision guide munitions, targeting lasers and satellite communications. Rare earths are produced almost exclusively in China, which raises concerns over potential supply disruptions.

High-performing alloys that are used, for instance, in fuselages of combat aircraft, require specific raw materials such as niobium, vanadium or molybdenum. Other alloys are based on titanium, which provides high specific strength and corrosion resistance, at just half the weight of steel and nickel-based superalloys. These properties make them indispensable in aeronautic applications. Beryllium is also used as a lightweight alloy in jet fighters, helicopters and satellites as it is six times lighter and stronger than steel, thus enables reducing weight and improving the speed and manoeuvrability. Beryllium finds also applications in missile gyroscopes, gimbals and for inner stage joining of elements in missile systems. Carbon fibres represent a key constituent of military aircraft, strategic missiles and satellites thanks to their superior stability, low coefficient of thermal expansions, high strength, high stiffness, low density and high abrasion resistance. However, only a few companies in Europe produce carbon fibres and their precursors. The same occurs for high purity gallium arsenide used in computer chips. Only a few producers are capable of producing compounds able to meet this high quality requirement, for which there are currently no effective substitutes.

Further developments in civil and defence technologies could have a big impact on the future demand for some raw materials, for instance lithium and cobalt in batteries, scandium in titanium-based alloys or neodymium and praseodymium in the production of NdFeB permanent magnets used in electric engines.

While access to raw materials is important for all European industries, an uninterrupted supply of materials to the defence industry is of particular importance, in view of its pivotal role in ensuring Europe's security. With the central objective of strengthening the defence industry, the European Commission stressed on the importance of raw materials for defence applications and raised concerns about their supply, which could hamper the competitiveness of the defence sector [1]. Although the raw materials used in the defence sector are often the same materials as in civil applications, a thorough assessment of type, quality and quantity of the demand in the European defence sector is currently not available, and at the same time very necessary to inform the debate amongst European policy makers.

The scope of this work is to identify which materials are important to the European defence industry and to assess their role in the manufacture of representative defence applications. This paper takes stock of our extended analysis on the inventory of raw materials used in major defence applications, which was previously published in a science for policy report [4], and further compares these materials with the latest list of CRMs for the EU economy.

2 Methodology and Approach

A top-down methodology was followed in this assessment. In a first step, the relevant European defence applications were identified from the land, air, naval, space, electronic and missile sectors, and subsequently disaggregated into subsystems and components. In a second step, the processed and raw materials used in the production of these subsystems and components were identified (Fig. 1).

The data collection began by reviewing the literature covering the materials required for the production of military equipment and defence products manufactured by the European defence industry. This literature review was followed by discussions by phone, questionnaires or interviews with industry researchers, specialised consultants and programme managers involved in R&D, product development and supply logistics. Overall inputs, information or validation of data used in this analysis have been received from representatives of the following organisations: CEIS, BIO Intelligence Services, Airbus Group, Meggitt Avionics, Rolls-Royce, Thales, Astrium, BAE Systems, CSM Italy, Safran, Snecma, and the French Ministry of Defence.

The methodology takes into account the constraints that are specific to the defence sector, such as the open literature offers limited information on the type of materials

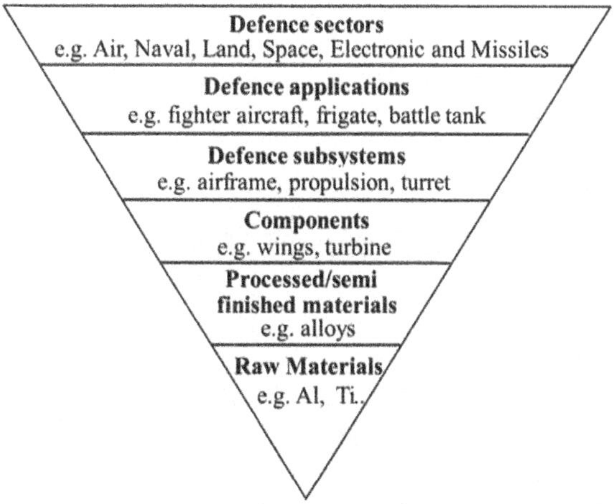

Fig. 1 Top-down approach used in this research

and their use in defence applications, and industrial stakeholders are reluctant in general to share information bound by confidentiality.

Once the data about the materials used in various defence applications were collected, they were analysed, extrapolated and the information gaps were overcome through approximation in order to obtain a representative picture of the European defence sector. The term 'important' material for defence sector is used to make reference to the capacity of these materials to fulfil the stringent requirements of defence applications, and for which there are no effective substitutes.

Finally, the raw materials that were identified as constituents of advanced processed materials used in the production of defence-related subsystems and components were compared with those that are included in the 2017 EU's critical raw materials list.

3 Results and Discussion

3.1 Overview of the European Defence Applications

The European defence industry comprises a large number of original equipment manufacturers (OEMs), also known as prime contractors, system builders, equipment suppliers and a complex network of suppliers. To build up the inventory of raw and processed materials used by the defence industry, we identified initially the representative defence applications produced by the European OEMs that feature on the SIPRI top 100 arms-producing companies in the world [5]. Three additional criteria were applied when selecting the materials required for the production of such defence products such as: (1) applications that require large quantities of materials to be built (e.g. warships) and those that are produced in high number (e.g. main battle tanks), (2) applications that are strategic assets (e.g. fighter aircraft), and (3) applications that are high-tech systems (e.g. satellites), requiring specific materials or of high purity. These applications are grouped into six sectors: air, naval, land, space, electronic and missile (Table 1).

3.2 Raw and Processed Materials in Defence Applications

Each defence application was split into subsystems and components, and the composition of those components in terms of materials has been identified. As an example, the list of materials used for components in the subsystems of an aircraft (fighter, transport, maritime patrol and unmanned) is presented in Table 2. The list on the decomposition of all defence applications into subsystems, components and materials is presented in our previous report [4].

Table 1 Representative European defence applications and system integrators

Defence application	Example	System integrator	Criteria
Air sector			
Fighter aircraft	Rafale Eurofighter Gripen F-35	Dassault Aviation Airbus D&S, Finmeccanica, BAE Saab Finmeccanica, BAE	2, 3
Combat helicopter	Tiger A129 Mangusta	Airbus helicopters AgustaWestland	2, 3
Transport aircraft (heavy)	A400M	Airbus D&S	1, 2
Transport aircraft (tactic)	CN-235	Airbus D&S	2
Maritime patrol aircraft	Falcon 50 M	Dassault Aviation	2, 3
Multi-role helicopter	NH90 AW101	Airbus Helicopters AgustaWestland	2, 3 2, 3
Unmanned aerial vehicle	S100 Camcopter	Schneibel	2, 3
Unmanned fighter aerial vehicle	nEUROn	Dassault aviation	2, 3
Naval sector			
Aircraft carrier	Elizabeth-class	BAE systems	1, 2, 3
Amphibious assault ship and helicopter carrier	Mistral-class Juan Carlos-Class	DCNS Navantia	1, 2, 3
Destroyer and frigate	FREMM Type 124	DCNS and Fincantieri TKMS	1, 2, 3
Corvette	Holland-class	Damen	1, 2, 3
Offshore patrol vessel	Gowind-class	DCNS	1, 2, 3
Submarine	Type-212 Scorpene-class	TKMS DCNS	1, 2, 3
Torpedo	Spearfish Black Shark	BAE systems Whitehead Sistemi Subacquei	2, 3
Land sector			
Main battle tank	Leopard-2	Krauss-Maffei Wegmann	1, 2
Infantry fighter vehicle	VBCI Puma IFV Boxer	Nexter KMW and Rheinmetall Rheinmetall	1, 2
Armoured personnel carrier	Patria AMV	Patria	1, 2
Self-propelled artillery	CAESAR Panzerhaubitze 2000	Nexter Krauss-Maffei Wegmann and Rheinmetall	1, 2
Towed artillery	105 LG	Nexter	1, 2

(continued)

Table 1 (continued)

Defence application	Example	System integrator	Criteria
Ammunitions	M982 Excalibur (155 mm) BONUS (155 mm) DM53 (120 mm)	BAE systems Bofors Nexter and BAE systems Bofors Rheinmetall	2
Assault rifle	G36	Heckler and Koch	2
Space sector			
Infrared satellite	Spirale	Thales (TAS)	2, 3
Radar satellite	SAR Lupe	OHB System	2, 3
ELINT satellite	CERES	Airbus D&S	2, 3
Communication satellite	Syracuse	Thales (TAS)	2, 3
Optical satellite	Helios	Airbus D&S, Thales Alenia Space	2, 3
Navigation satellite	Galileo	Airbus D&S, Thales Alenia Space	2, 3
Launch systems	Ariane-V Vega	Airbus D&S ESA and Italian Space Agency	1, 2
Electronic sector			
Radar	Giraffe	Saab	2, 3
Communication systems	FIST	Thales UK	2, 3
Missile sector			
Air-air missile	MICA IR	MBDA	2, 3
Surface-air missile	Aster 30	MBDA	2, 3
Anti-tank missile	MILAN	MBDA	2, 3
Anti-ship missile	Exocet	MBDA	2, 3
Cruise missile	SCALP/Storm shadow	MBDA	2, 3

The analysis showed that similar subsystems, for example, the 'airframe' exists in different air sector applications such as in transport, fighter and unmanned aircraft, and they comprise the same components (e.g. 'body', 'wings', 'tail', 'nose' and 'axis'). The components for these similar subsystems are made mostly from the same types of material, although in different quantities.

It is worth to mention that while information about the composition of the components and subsystems was readily obtained for the air, naval and land sector's applications, in certain cases (e.g. the space and electronic sectors) this information was not found, for two main reasons:

- The information was not shared by stakeholders due to issues of confidentiality, either from commercial or security perspectives;
- The lead system integrators or even the subcontractors downstream in the supply chain are not fully aware or do not know which materials are used in their systems.

Table 2 Decomposition of an aircraft (fighter, transport, maritime patrol and unmanned) into subsystems, components and materials

Subsystem	Component	Materials used
Airframe	Body	Carbon-epoxy composite (fibre T800) Aluminium alloy 7000 series (7010, 7040, 7050, 7075) and 2000 series (e.g. 2024) Other alloys containing: beryllium, cadmium, chromium, germanium, gold, lead, nickel, tantalum, tin, tungsten and molybdenum
	Wings	Carbon-epoxy composite (fibre T800) Aluminium alloys 2000 series (e.g. 2024) and 7000 series (e.g. for upper skin) Titanium alloy, mainly TA6V
	Tail	Carbon-epoxy composite (fibre T800) Aluminium alloy 2000 series (e.g. 2024) Titanium alloy, mainly TA6V
	Nose	Kevlar, glass-fibre-epoxy
	Axis	Special steel alloy, e.g. MARVAL X12
Propulsion	Fan and compressors	Mainly titanium-based alloys (e.g. TA6V) Carbon-fibre composites
	Combustor	Nickel superalloy (e.g. N-18) Aluminium-magnesium alloy in the carter Superalloys (e.g. HS25, HS31, HS188)
	Turbine (high pressure and lower pressure)	Nickel single-crystal alloy (e.g. AM1) Nickel-aluminium-platinum or yttrium-stabilised zirconia in the coating of turbine blades (thermal barrier coating) Nickel alloy (e.g. Inco 718)
	Nozzle	Carbon-carbon composite
	Post combustion	Carbon-carbon composite
	Drive shaft and propellers (for maritime patrol and transport aircraft)	Kevlar composite

(continued)

Table 2 (continued)

Subsystem	Component	Materials used
Landing gear	Landing gear body	Aluminium alloys Copper-beryllium alloys Titanium and titanium alloys Special steels, e.g. MLX19
	Wheels	Rubber
	Brakes	Composites
Electronic systems	Sensors and avionics	Gallium arsenide Mercury-cadmium-telluride crystals Germanium and compounds with germanium Kevlar Aluminium alloys Titanium alloys Other compounds, substrates and alloys containing: rare earths, nickel, boron, indium and cobalt
	Communication/identification systems	Gallium arsenide 99.999+% Gallium nitride 99.999+% Beryllium alloys (copper-beryllium, aluminium-beryllium) Other compounds containing silver
	Energy storage	Barium titanate Lithium compounds
	Electro-optical systems	Mercury cadmium telluride crystals Neodymium-doped yttrium aluminium garnet Kevlar Copper-beryllium alloys and beryllium oxides Carbon-carbon composites Rare earths (all except scandium, promethium, terbium, lutetium) Gallium arsenide Other compounds with indium, selenium, germanium or boron Zinc sulphide Zinc selenite Thorium fluoride Tantalum peroxide Hafnium dioxide Titanium dioxide

(continued)

Table 2 (continued)

Subsystem	Component	Materials used
Weapon systems	Canon, missile and bomb (for fighter aircraft)	N.A.
	Chaff (counter measure)	N.A.
	Flares	N.A.
Connectors	All integrated wire network	Aluminium-beryllium alloys Copper
Canopy	Canopy	Special glass Germanium compounds

Industrial stakeholders admitted that materials flow becomes very opaque a couple of tiers down in the supply chain.

The analysis shows that the defence equipment is mainly made from processed and semi-finished materials, for example, composites, metallic alloys, etc. Various composition of alloys are used in the production of many defence applications, such as: aluminium-based alloys (2000 and 7000 series) and titanium-based alloy TA6V (90% titanium, 6% aluminium and 4% vanadium) in the airframe of a fighter aircraft; superalloys containing nickel with single crystal microstructure (e.g. N-18 and AM1) for blades in the turbines of jet fighters; aluminium, titanium and steel alloys in various equipment working in a corrosive high-temperature environment. In naval applications, the hull of submarines is made of ultra-high strength steel alloy (HSLA) containing nickel, molybdenum, manganese and vanadium. Cannon barrel and missile bodies are also made of special steels (e.g. Marval X12) and nickel alloys. Aluminium alloys are used to manufacture torpedoes. Another example is gallium, which is used as gallium arsenide, gallium nitride or gallium antimonide in communications, night vision, radar and satellites.

Advanced composites (e.g. carbon and aramid fibre composites) are progressively replacing aluminium alloys in many types of equipment, in particular in aerodynamic applications since they are able to meet the high-performance standards. Such composites are extremely strong and at the same time lighter than aluminium alloys. In fact, the composite materials are used in a higher amount in the recent types of jet fighters (Fig. 2).

It is important to mention that most prime contractors and OEMs do not buy and process themselves the raw materials, but rather acquire semi-finished alloys and other compounds or even the actual components.

Overall we found that 47 different types of alloys, composites and compounds are mostly used for the manufacture of defence application systems (Table 3).

When looking at the raw materials as the constituents of the alloys and compounds presented in Table 3, we identified that 39 raw materials are mostly necessary for their production, and therefore for the manufacture of defence-related subsystems and components. Based on their chemical structure and properties, these raw materials

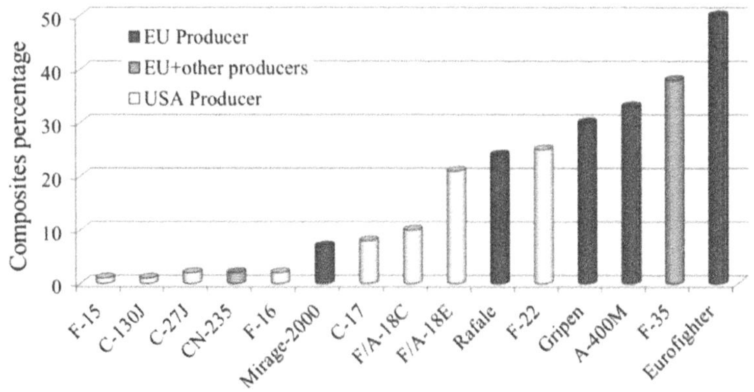

Fig. 2 Estimated composites loading in various defence aeronautic applications [4]

are grouped into four categories: metals, precious metals, rare earth elements and non-metals (Fig. 3).

3.3 Potential Supply Issues of the Raw and Processed Materials Used in Defence Applications

Although the demand for raw materials used in the production of defence applications is relatively low, some of them are subject of concerns over their security of supply. A particular challenge of the European defence industry is related to the downstream supply of processed materials, including materials processing know-how and transformation capabilities. The EU is a large manufacturer of alloys and special steel, but has limited production capacities for speciality composite materials and their precursors.

Many defence applications require the same materials that are also used in the civil sector (dual-use applications). However, in some cases the defence sector needs special steels or alloys. Another difference to 'civil' materials is that the defence applications require higher purity or special composition of the alloys.

When taking into account the EU's import dependency on raw materials used by the EU defence industry, it is revealed that the EU is fully dependent on imports of 13 of the 39 raw materials (i.e. boron (as borates), dysprosium, gold, magnesium, molybdenum, neodymium, niobium, praseodymium, samarium, tantalum, titanium, yttrium and other REEs). Overall, for over two-thirds of those raw materials, the share of imports exceeds 50% (Fig. 4a). Due to data limitation, this figure does not take into consideration the import reliance of four raw materials, such as beryllium, cadmium, thorium and zirconium.

China is the major global producer of 58% of raw materials identified as important for defence applications (Fig. 4b). The supply risk of raw materials produced in

Table 3 List of relevant processed materials used in defence applications

Processed materials	Typical component/subsystem
Alloys	
Aluminium alloy 2000 and 7000 series (2024, 7010, 7040, 7050)	Internal structural parts of aircraft fuselage and panelling
Aluminium-beryllium alloys	Electronic systems for communication and identification
AM1 Nickel mono-crystalline alloy	Jet turbines in propulsion systems
Bronze	Bushes, bearings, airframe tail and propulsion systems in azimuth thrusters
Cadmium mercury telluride	Electro-optical systems for sensors and infrared detectors
Copper-beryllium alloys	Landing gear body and on-board electronics for phased array radar
HS25, HS31, HS188 cobalt alloys	Combustor components used in propulsion
Inconel and other nickel alloys (e.g. Waspaloy, Nimonic, Udimet, etc.)	Refractory alloy used in aircraft turbojet engines for discs, blades and casing; hangar and desks' super-structure of aircraft and frigates, and for high-strength bolts
Light aluminium alloys (other than 2000 and 7000 series), e.g. 5059, 6061 and others (e.g. aluminium-copper-lithium alloys)	Engineering structures and components of lightweight defence vehicles (including marine); armour structures, rocket, missile casing, pipe systems such as fuel, frames and structural parts
Light alloys—others	Lightweight armour vehicles, rockets, missiles, marine and aircraft structures
N-18 Nickel superalloy	Turbines, fan and compressors in aircraft engines
Single crystal materials (e.g. CMSX4 alloy)	Major component of turbine engines for aeronautics applications (e.g. helicopters)
Special stainless steels (e.g. MLX12 and MLX12H)	Barrel structure of assault rifle
Special steel alloy MARVALX12	Axis of ailerons and propulsion arming and firing
Special steels MLX19	Landing gears in aircraft and helicopter applications
Special steels (e.g. NCAV, FDMA, 819B, MY18, FADH, 819AW, GHK, FND, GH4 W, F65)	Structure of assault rifle body
Special steels MARVAL X12H, MARAGING 250 & 300, APX4, ALUMINUM, 17.4Ph, 15.5Ph	Cruise and propulsion motor of missiles

(continued)

Table 3 (continued)

Processed materials	Typical component/subsystem
Steel and corrosion resistant steel (CRES steels)	Armour plate for tanks, field artillery pieces; components used in aircraft carriers and submarines; fasteners, bolts
Titanium alloy: TA6V (Ti-6Al-4V), $Ti_3Al_{2.5}V$	Tail and wings of airframes, landing gears and propulsion combustor; high-loaded and fatigue-critical parts; hydraulic pipes
Very high hardness steel	Armour components, armour plate, armour piercing projectiles
Composites	
Carbon laminate	Drive shaft (frames and beams)
Carbon composite	On-board electronic for bow sonar and periscope; seven-bladed skewback propeller; super-structure of desks and body frame
Carbon-carbon composite	Post-combustion parts; propulsion systems; nozzles
Carbon-epoxy composite	Aircraft structural parts (airframe, wings, fuselage)
Ceramic-carbon fibres	Structural parts of missile (e.g. nose cap); heat-shield material for emergency systems/flight recorders on fighter aircraft (black box)
Co-fired ceramics (low and high temperature)	Optoelectronic module and sensor packaging
Composite materials (carbon reinforced polymer fibres and Kevlar)	Rotor blades, fuselage, drive shafts, flight controls and transmission; solar arrays for power supply and antenna of satellites
Composites—other	Various components of aircraft, tank armour and liners, fighter vehicles, missiles, superstructures, turrets, propeller and parts of modern submarines
Fibre-composite	Rotors for dynamic parts of helicopter
Fibre elastomer	Drive shaft for propulsion
Glass fibre	Components of assault rifle and guided missile motor casings
Glass-fibre-reinforced plastics	Rotors for dynamic parts of helicopter
Nomex honeycomb with carbon and Kevlar	Canopy and satellite structures
Compounds	
Ceramics	Armour components and combustion chambers of propulsion systems
Epoxy resin	Guidance package of missile's radome
Fused silica ceramics	Sensors/seeker of missile's radome

(continued)

Table 3 (continued)

Processed materials	Typical component/subsystem
Gallium arsenide	Fibre-optic data communications, wireless network, modern night-vision technology, phased array radar, sensor and avionics, other electro-optical systems and satellite (highly efficient solar cells)
Gallium nitrite	Semiconductor for electronic devices
Glass	Optical devices; as sapphire glass in armour
Kevlar	Structural part of airframe, drive shaft, canopy, body armour and flak jackets
Neodymium Yttrium-Aluminium garnet (Nd-YAG)	Lasers (long-range finders and target designators), phased array radar
Piezoelectric ceramics	Active element for sonar
Polyamide resin	Body frame of missile
Quartz ceramics	GPS/SAL guidance system, inertial navigation, combat identification equipment/IR
Rubber	Tyred road wheels for running gears, wheels
Silicon rubber	Sensors/seeker of missile's radome
Special glass	Canopy

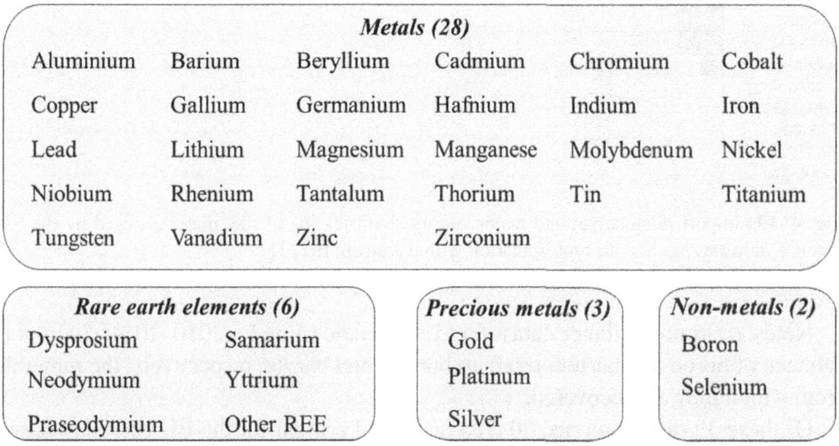

Fig. 3 Relevant raw materials used in defence applications

this country is considered high and it may be interrupted (e.g. by imposing export restriction and taxes). Among these 39 raw materials, the EU is the largest global supplier only for hafnium.

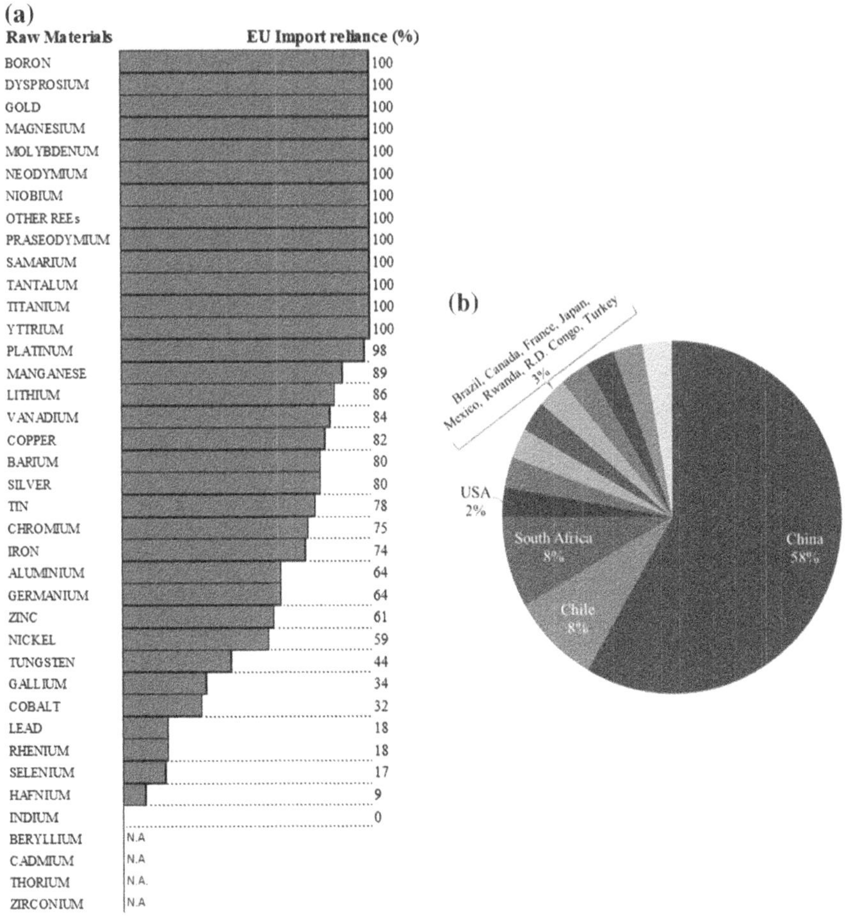

Fig. 4 EU import reliance (**a**) and major supply countries (**b**) of raw materials used by the EU defence industry. *Source* our representation with data from EC [3]

Notes: (i) import reliance data refer to an average value for 2010–2014; (ii) import reliance of boron and barium refers to borates and baryte, respectively, the minerals from which they are recovered.

Of these 39 raw materials, 20 are considered critical for the EU economy based on the latest assessment published by the European Commission in 2017 [2]. They are: barium, beryllium, boron, cobalt, dysprosium, gallium, germanium, hafnium, indium, magnesium, neodymium, niobium, platinum, praseodymium, samarium, tantalum, tungsten, vanadium, yttrium and other REEs.

The position of these 39 raw materials in the EU criticality matrix, which is represented by the plot of economic importance and supply risk, is presented in Fig. 5.

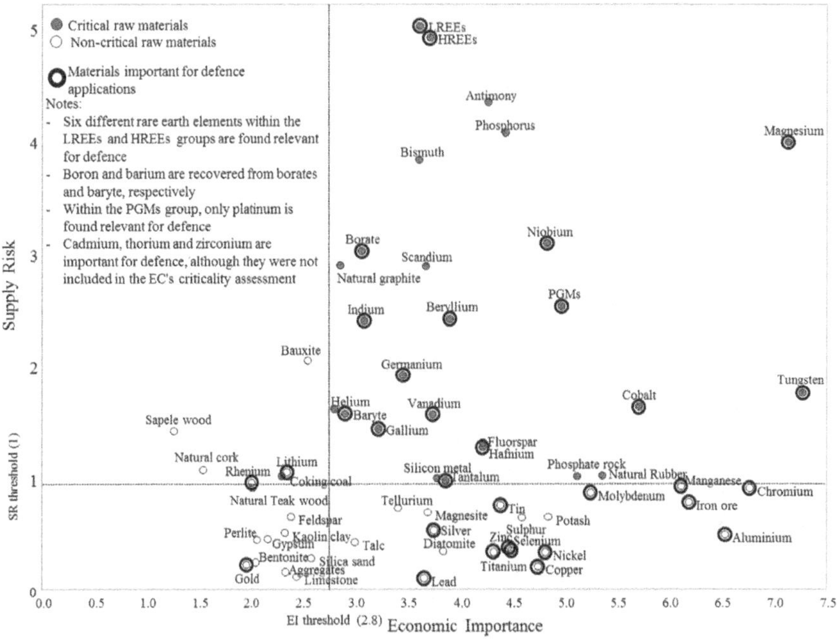

Fig. 5 Raw materials in defence industries in relation to the 2017 EU criticality matrix

Based on the criticality rating assigned to these 20 raw materials and their use in the specific defence subsectors, it is indicated that the aeronautics and electronics are the most vulnerable to potential materials supply constraints. In view of the strategic importance of the defence sector for Europe's security and the need to increase its competitiveness, it is imperative that the related manufacturing industries operate under uninterrupted conditions. Therefore, the European defence industry needs to secure the supply of a number of raw materials from international sources, maintain its global leadership in the manufacture of high-performance alloys and special steel, and further develop capabilities for the production of speciality composite materials to tackle the supply risks associated with raw and processed materials used in the defence sector.

4 Conclusions

An assessment of the most representative applications of the European defence industry revealed that 47 different alloys, composites and compounds are important for the manufacture of the European defence application systems. 39 raw materials are mainly necessary to manufacture such advanced materials, of which about half are considered critical for the EU economy. The EU is totally depended on imports for 13 of the 39 raw materials and for over two-thirds of them the share of imports exceeds 50%.

Given the very high level of performance and special properties of the processed materials that cannot be matched by readily available substitutes, their potential supply risk is perceived by the defence industry as being even much higher compared to that of the constituent raw materials (e.g. minerals and metals).

A set of actions, such as strengthening the materials supply chain, materials processing know-how and materials transformation capabilities, is needed to be taken for improving the security of supply of processed and raw materials used in the European defence industry. In particular, the EU will need to maintain its leadership in the manufacture of alloys and special steels, and to improve its production capacities for specialty composite materials (e.g. aerospace-grade carbon fibres and their precursors).

References

1. EC (2013) European Commission's Communication. Towards a more competitive and efficient defence and security sector. COM(2013) 542, 2013, Brussels
2. EC (2017a) European Commission's Communication. On the 2017 list of critical raw materials for the EU. COM(2017) 490, 2017, Brussels
3. EC (2017b) European Commission, Study on the review of the list of critical raw materials—Criticality assessments, June 2017
4. Pavel CC, Tzimas E (2016) Raw materials in the European defence industry, EUR 27542 EN, https://doi.org/10.2790/509931
5. SIPRI (2017) Stockholm International Peace Research Institute (SIPRI), SIPRI Arms Industry Database. https://www.sipri.org/

Molecularly Imprinted Polymers and Optical Fiber Sensors for Security Applications

Nunzio Cennamo, Maria Pesavento, Simone Marchetti and Luigi Zeni

Abstract In recent years several approaches to the optical monitoring of Molecularly Imprinted Polymers (MIPs) via optical fibers have been developed with very promising developments in the field of sensing. In this paper, we report about sensing platforms based on MIPs combined with Plastic Optical Fiber (POF) which appear to be useful for security and defense. In particular, two optical chemical sensor configurations for 2,4,6-trinitrotoluene (TNT) detection in water are reported. These sensor configurations are based on the same Molecularly Imprinted Polymer receptor, but the first one is combined with a Surface Plasmon Resonance (SPR) platform and the second one with a Localized Surface Plasmon Resonance (LSPR) platform, both realized exploiting D-shaped Plastic Optical Fibers (POFs).

Keywords Optical sensors · Chemical sensors · Plastic optical fibers · Molecularly imprinted polymers · 2,4,6-trinitrotoluene (TNT) · Plasmonic sensors

1 Introduction

In security and defence analytical controls, where a selective detection of dangerous substances is necessary, the biosensors technology based on selective receptors intimately connected to a transducer, is particularly suitable. More specifically, the optical fiber biosensors are well known for their advantages when used as intrinsic or extrinsic optical devices for sensing [1–7]. When the fiber interacts with the analyzed medium directly, the system can be classified as intrinsic optical fiber sensor, whereas when the fiber is used only as waveguide for the propagation of the light in the sensing region and for collection, it can be defined as extrinsic optical fiber sensor. Moreover, in optical fiber sensors can be used two different detection schemes, reflection or transmission mode, with several with several transduction mechanisms.

N. Cennamo (✉) · L. Zeni
Department of Engineering, University of Campania Luigi Vanvitelli, Aversa, Italy
e-mail: nunzio.cennamo@unicampania.it

M. Pesavento · S. Marchetti
Department of Chemistry, University of Pavia, Pavia, Italy

R. Fangueiro and S. Rana (eds.), *Advanced Materials for Defense*, 4, Springer Proceedings in Materials, https://doi.org/10.1007/978-3-030-34123-7_2

In reflection mode the light source and the detector are placed on the same side of the fiber, whereas in transmission mode they are placed on opposite sides.

In the last years, several silica optical fiber biosensors based on plasmonic phenomenon, surface plasmon resonance (SPR) or localized surface plasmon resonance (LSPR), have been presented by several research groups [7–22], e.g. Homola et al. [8, 9], Gupta et al. [10, 11] and Albert et al. [12]. SPR is widely used as a detection principle for many sensors operating in different application fields, such as bio and chemical sensing, mainly because this transduction technique is suitable for marker free sensors.

When artificial receptors are used for bio/chemicals detection, the thin dielectric layer on the metal surface (usually a gold surface) selectively recognizes and captures the analyte present in a liquid sample, so producing a local change in the refractive index at the metal surface. The value of the refractive index variation depends on the structure of the analyte molecule and on the amount of molecules interacting with the receptor sites [7–10].

Recently, several research groups have realized optical fiber sensors exploiting Plastic Optical Fibers (POFs), thus providing an effective approach to low-cost sensing. POFs are used for their advantages in addition to those of optical fibers, such as immunity to electromagnetic fields, remote sensing, small size and weight. These low-cost POF sensor platforms can be used combined with artificial receptors (i.e. MIPs) or bio-receptors for selective detection of substances in different application fields [23, 24]. For instance, a review on POF sensors based on light intensity variation is presented by Bilro et al. [23], whereas recent progresses in POF sensors are presented by Jin and Granville [24]. F. Sequeira et al. have presented a simple and highly sensitive D-shaped POF platform for evanescent field sensing, through an intensity-based configuration [25]. N. Cennamo et al. have presented a plasmonic POF sensor based on a multilayer deposited on a D-shaped POF area [20, 26]. Moreover, in the last years Cennamo et al. have implemented important applications based on this plasmonic POF platform with gold film and different kinds of receptors, such as MIPs and bio-receptors [27–35]. The authors have also presented MIPs combined with different POF platforms, i.e. based on tapered POFs [31], gold nanostars dispersed in MIP [32] and a trench milled between two POFs filled with MIP [33]. In this work we explain two chemical (marker free) sensors, focused on technical novelties of interest in the field of defense, based on MIP receptor combined with plasmonic POF platforms.

To obtain good chemical stability and selectivity the molecular imprinting technique can be used for the preparation of molecular-recognition materials, with many favorable aspects over bio-receptors; such as an easier and faster preparation, the possibility of application outside the laboratory, for example in real environmental conditions, a longer durability [36–38]. Moreover, MIPs can be deposited in a simple and reproducible way by spinning the pre-polymeric mixture on the flat sensing surface of the D-shaped fiber and by polymerizing in oven [30–32, 34].

A specific MIP for 2,4,6-trinitrotoluene (TNT) detection in water [30, 32], monitored by two POF sensing devices, is presented in this work. The detection of explosives in aqueous samples is highly required for different reasons. Thus, because of

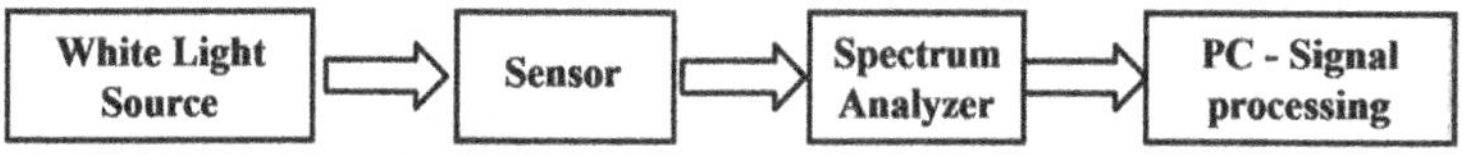

Fig. 1 Outline of the experimental setup

the need of a rapid on-site and remote detection of explosives in aqueous samples, several sensors have been proposed [39–42].

2 Experimental Setup

In this work, in order to measure the light spectrum transmitted through the sensors, the same experimental setup for both the TNT sensor configurations has been used. It consists of a halogen lamp, illuminating the biosensor system, and a spectrum analyzer [20, 30, 32]. Figure 1 shows the outline of this simple and low-cost experimental setup.

The employed halogen lamp exhibits a wavelength emission range from 360 to 1700 nm, while the spectrometer detection range goes from about 330 to 1100 nm.

3 An SPR-MIP TNT Sensor

Figure 2 shows the plasmonic POF platform used to monitor the refractive index variations of an MIP receptor upon combination of TNT. The surface plasmon resonance phenomenon is obtained in a D–shaped POF area covered with an optical overlayer between the exposed POF core and the top thin gold film [20]. More specifically, to make this platform we removed the cladding and part of the core by two different polishing papers, spun the buffer layer (Microposit S1813, Shipley Company, Marlborough, MA) on the exposed core and, at the end, a thin gold film was sputtered.

The used POF has a core of polymethyl methacrylate (PMMA) with a diameter of 980 μm and a cladding of fluorinated polymer. The total diameter of the POF is 1 mm, the D-shaped POF region presents a length of 1 cm with a width of about 850 μm, whereas the multilayer has a buffer layer of about 1.5 μm thick covered by a gold film with a thickness of 60 nm. The optical buffer layer is used to improve the performances of the SPR sensor, since it increases the interaction of the evanescent wave with the surface plasmons of the gold layer, given that the buffer layer presents a higher refractive index than the one of the POF core [20].

Figure 2 shows the SPR platform, previously described, with a planar gold surface covered by the MIP receptor layer, used to obtain the selective detection of the TNT [30]. Before depositing the MIP receptor, the gold planar sensing area was washed with ethanol. After that, it was dried in a thermostatic oven at 60 °C prior to deposit the

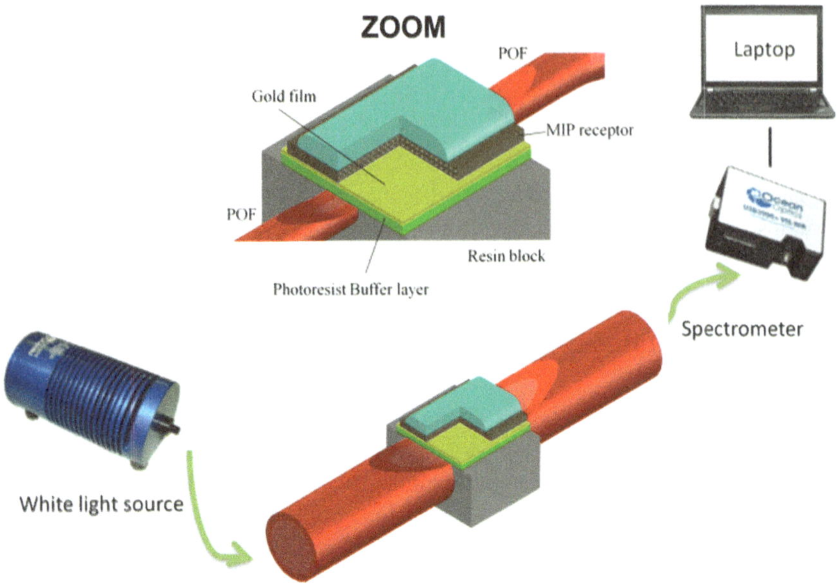

Fig. 2 SPR-MIP sensor in a D-shaped POF

polymer layer (MIP). In particular, about 100 μL of the prepolymeric mixture were dropped over the gold surface and spun for 35 s at 1000 rpm. The polymerization step was then carried out for 16 h at 70 °C in oven. Finally, the obtained polymeric film was washed and the template molecule (TNT) was extracted, leaving the imprinting sites free for rebinding [30].

4 An LSPR-MIP TNT Sensor

In this section, we report a high sensitive optical chemical sensor based on gold nanostars [43], MIP receptor [30], and tapered POF [31, 32]. In the following we briefly explain how this sensor can be produced. The POF (the same POF used in Sect. 3) was heated at 150 °C and stretched with a motorized linear positioning stage (Physik Instrumente (PI) M150.11). After this step, in order to remove the cladding, the POF was POF was embedded in a resin block and polished as described in Sect. 3. The modified region (D-shaped tapered POF) was about 1 cm in length. Finally, the sensing layer, an MIP with suspended five-branched gold nanostars (GNS) [32, 43], was prepared and deposited on the exposed core of the D-shaped tapered-POF, as described in [32]. Figure 3 shows this special tapered-POF platform covered with the GNS-MIP layer.

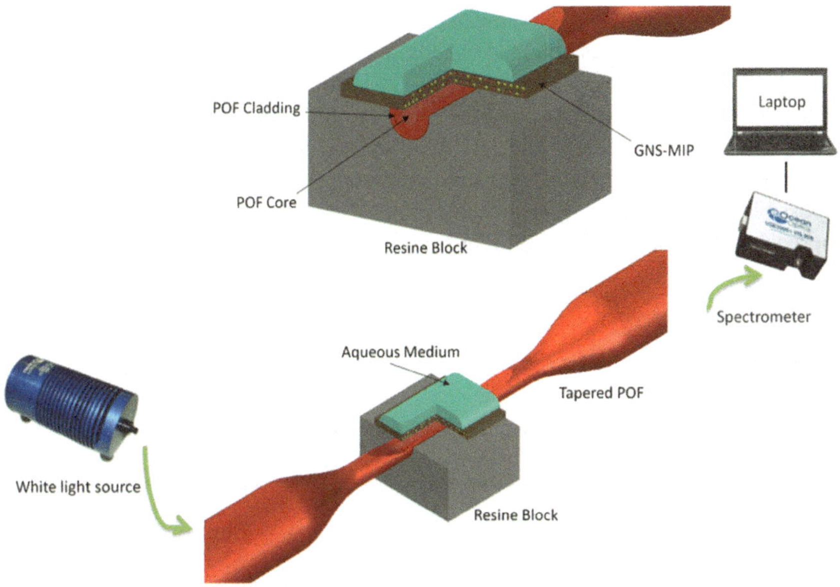

Fig. 3 LSPR-MIP sensor in tapered D-shaped POF

5 Experimental Results and Conclusions

Two different plasmonic POF platforms combined with MIPs for TNT detection are reported. In the first case (Sect. 3), an MIP receptor for TNT is deposited on an SPR-POF platform [30], in the second one (Sect. 4), an MIP loaded with gold nanostars (GNS) is deposited on a D-shaped tapered POF platform (in this case, LSPR is the transduction method) [32].

The resonance wavelength variation with the TNT concentration is in agreement with the Langmuir absorption isotherm. This relation can be used for the analytical determination of TNT by the proposed sensors [30, 32].

The parameters of the Langmuir model, i.e. the sensitivity at low concentration, and the limit of detection (LOD), are summarized in Table 1 [30, 32], together with the affinity constant of the receptor site, i.e. the equilibrium constant of the association of TNT with the specific sites in MIP [32].

Table 1 Performances of TNT sensors in POFs [32]

TNT sensor in POFs	Sensitivity (nm/M)	LOD (M)	K_f (M^{-1})
MIP on gold layer/POF	2.7×10^4	5.1×10^{-5}	4×10^4
GNS-MIP on tapered POF	8.3×10^5	7.2×10^{-7}	4×10^4

It is interesting that K_f is the same in the two configurations, as it is expected since the receptor, i.e. MIP, was obtained exactly in the same way [32]. The sensitivity of GNS-MIP on tapered POF configuration [32] is about thirty times higher than that of the SPR-POF sensor with an MIP layer on the gold film [30]. We have also demonstrated in [32] the good selectivity of the sensing devices here proposed.

The proposed POF-MIP sensing configurations are easy to implement and make it possible the detection at low concentration. They can pave the way toward a new approach in the chemical optical fiber sensors for security applications.

References

1. Leung A, Shankar PM, Mutharasan R (2007) A review of fiber-optic biosensors. Sens Act B Chem 125:688–703
2. Wang XD, Wolfbeis OS (2013) Fiber-optic chemical sensors and biosensors (2008–2012). Anal Chem 85:487–508
3. Wang XD, Wolfbeis OS (2016) Fiber-optic chemical sensors and biosensors (2013–2015). Anal Chem 88:203–227
4. Monk DJ, Walt DR (2004) Optical fiber-based biosensors. Anal Bioanal Chem 379:931–945
5. Baldini F, Brenci M, Chiavaioli F, Giannetti A, Trono C (2012) Optical fibre gratings as tools for chemical and biochemical sensing. Anal Bioanal Chem 402:109
6. Consales M, Ricciardi A, Crescitelli A, Esposito E, Cutolo A, Cusano A (2012) Lab-on-fiber technology: toward multifunctional optical nanoprobes. ACS Nano 6:3163
7. Jorgenson RC, Yee SS (1993) A fiber-optic chemical sensor based on surface plasmon resonance. Sens Act B Chem 12:213–220
8. Homola J (2003) Present and future of surface plasmon resonance biosensors. Anal Bioanal Chem 377:528–539
9. Homola J, Yee SS, Gauglitz G (1999) Surface plasmon resonance sensors: review. Sens Act B Chem 54:3–15
10. Anuj K, Sharma RJ, Gupta BD (2007) Fiber-optic sensors based on surface Plasmon resonance: a comprehensive review. IEEE Sens J 7:1118–1129
11. Gupta BD, Kant R (2018) Recent advances in surface plasmon resonance based fiber optic chemical and biosensors utilizing bulk and nanostructures. Opt Laser Technol 101:144–161
12. Caucheteur C, Guo T, Albert J (2015) Review of plasmonic fiber optic biochemical sensors: improving the limit of detection. Anal Bioanal Chem 407:3883–3897
13. Sanders M, Lin Y, Wei J, Bono T, Lindquist RG (2014) An enhanced LSPR fiber-optic nanoprobe for ultrasensitive detection of protein biomarkers. Biosens Bioelectron 61:95–101
14. Mayer KM, Hafner JH (2011) Localized surface plasmon resonance sensors. Chem Rev 111:3828–3857
15. Petryayeva E, Krull UJ (2011) Localized surface plasmon resonance: nanostructures, bioassays and biosensing—a review. Anal Chim Acta 706:8–24
16. Sai VVR, Kundu T, Mukherji S (2009) Novel U-bent fiber optic probe for localized surface plasmon resonance based biosensor. Biosens Bioelectron 24:2804
17. Estevez MC, Otte MA, Sepulveda B, Lechuga LM (2014) Trends and challenges of refractometric nanoplasmonic biosensors: a review. Anal Chim Acta 806:55–73
18. Klantsataya E, Jia P, Ebendorff-Heidepriem H, Monro TM, François A (2017) Plasmonic fiber optic refractometric sensors: from conventional architectures to recent design trends. Sensors 17:12
19. Trouillet A, Ronot-Trioli C, Veillas C, Gagnaire H (1996) Chemical sensing by surface plasmon resonance in a multimode optical fiber. Pure Appl Opt 5:227–237

20. Cennamo N, Massarotti D, Conte L, Zeni L (2011) Low cost sensors based on SPR in a plastic optical fiber for biosensor implementation. Sensors 11:11752–11760
21. Emiliyanov G, Høiby PE, Pedersen LH, Bang O (2013) Selective serial multi-antibody biosensing with TOPAS microstructured polymer optical fibers. Sensors 13:3242–3251
22. Yuan W, Khan L, Webb DJ, Kalli K, Rasmussen HK, Stefani A, Bang O (2011) Humidity insensitive TOPAS polymer fiber Bragg grating sensor. Opt Express 19:19731–19739
23. Bilro L, Alberto N, Pinto JL, Nogueira R (2012) Optical sensors based on plastic fibers. Sensors 12:12184–12207
24. Jin Y, Granville AM (2016) Polymer fiber optic sensors—a mini review of their synthesis and applications. J Biosens Bioelectron 7:1–11
25. Sequeira F, Duarte D, Bilro L, Rudnitskaya A, Pesavento M, Zeni L, Cennamo N (2016) Refractive index sensing with D-shaped plastic optical fibers for chemical and biochemical applications. Sensors 16:2119
26. Cennamo N, D'Agostino G, Donà A, Dacarro G, Pallavicini P, Pesavento M, Zeni L (2013) Localized surface plasmon resonance with five-branched gold nanostars in a plastic optical fiber for bio-chemical sensor implementation. Sensors 13:14676–14686
27. Cennamo N, Alberti G, Pesavento M, D'Agostino G, Quattrini F, Biesuz R, Zeni L (2014) A simple small size and low cost sensor based on Surface Plasmon Resonance for selective detection of Fe(III). Sensors 14:4657–4671
28. Aray A, Chiavaioli F, Arjmand M, Trono C, Tombelli S, Giannetti A, Cennamo N, Soltanolkotabi M, Zeni L, Baldini F (2016) SPR-based plastic optical fibre biosensor for the detection of C-reactive protein in serum. J Biophotonics 9:1077–1084
29. Cennamo N, Zeni L, Tortora P, Regonesi ME, Giusti A, Staiano M, D'Auria S, Varriale A (2018) A high sensitivity biosensor to detect the presence of perfluorinated compounds in environment. Talanta 178:955–961
30. Cennamo N, D'Agostino G, Galatus R, Bibbò L, Pesavento M, Zeni L (2013) Sensors based on surface plasmon resonance in a plastic optical fiber for the detection of trinitrotoluene. Sens Actuators B Chem 188:221–226
31. Cennamo N, D'Agostino G, Pesavento M, Zeni L (2014) High selectivity and sensitivity sensor based on MIP and SPR in tapered plastic optical fibers for the detection of L-nicotine. Sens Actuators B Chem 191:529–5365
32. Cennamo N, Donà A, Pallavicini P, D'Agostino G, Dacarro G, Zeni L, Pesavento M (2015) Sensitive detection of 2,4,6-trinitrotoluene by tridimensional monitoring of molecularly imprinted polymer with optical fiber and five-branched gold nanostars. Sens Actuators B Chem 208:291–298
33. Cennamo N, Testa G, Marchetti S, De Maria L, Bernini R, Zeni L, Pesavento M (2017) Intensity-based plastic optical fiber sensor with molecularly imprinted polymer sensitive layer. Sens Actuators B Chem 241:534–540
34. Cennamo N, De Maria L, Chemelli C, Profumo A, Zeni L, Pesavento M (2016) Markers detection in transformer oil by plasmonic chemical sensor system based on POF and MIPs. IEEE Sens J 16:7663–7670
35. Cennamo N, Pesavento M, Lunelli L, Vanzetti L, Pederzolli C, Zeni L, Pasquardini L (2015) An easy way to realize SPR aptasensor: a multimode plastic optical fiber platform for cancer biomarkers detection. Talanta 140:88–95
36. Sellergren B (1997) Noncovalent molecular imprinting: antibody-like molecular recognition in polymeric network materials. Trends Anal Chem 16:310–320
37. Mayes AG, Mosbach K (1997) Molecularly imprinted polymers: useful materials for analytical chemistry? Trends Anal Chem 16:321–332
38. Haupt K, Mosbach K (2000) Molecularly imprinted polymers and their use in biomimetic sensors. Chem Rev 100:2495–2504
39. Smith RG, D'Souza N, Nicklin S (2008) A review of biosensors and biologically inspired systems for explosives detection. Analyst 133:571–584
40. Cerruti M, Jaworski J, Raorane D, Zueger C, Varadarajan J, Carraro C, Lee SW, Maboudian R, Majumdar R (2009) Polymer-oligopeptide composite coating for selective detection of explosives in water. Anal Chem 81:4192–4199

41. Giannetto M, Maialini E, Ferri EN, Girotti EN, Mori G, Careri M (2013) Competitive amperometric immunosensor based on covalent linking of a protein conjugate to dendrimer-functionalised nanogold substrate for the determination of 2,4,6-trinitrotoluene. Anal Bioanal Chem 405:737–743
42. Pesavento M, D'Agostino G, Alberti G, Biesuz R, Merli D (2013) Voltammetric plat-form for detection of 2,4,6-trinitrotoluene based on a molecularly imprinted polymer. Anal Bioanal Chem 405:3559–3570
43. Pallavicini P, Donà A, Casu A, Chirico G, Collini M, Dacarro G, Falqui A, Milanese C, Sironi L, Taglietti A (2013) Triton X-100 for three-plasmon gold nanostars with two photothermally active NIR (near IR) and SWIR (short-wavelength IR) channels. Chem Commun 49:6265–6267

Lanthanum Hydroxide Nanoparticles/Multi-Wall Carbon Nanotubes Nanocomposites

Arvind K. Bhakta, Sunita Kumari, Sahid Hussain, Ronald J. Mascarenhas, Praveen Martis, Joseph Delhalle and Zineb Mekhalif

Abstract A simple and effective approach using infrared (IR) irradiation and diazonium chemistry has been applied to decorate multi-wall carbon nanotubes (MWCNTs) with lanthanum hydroxide nanoparticles (LaH NPs). The decoration process comprised four steps: purification, functionalization, impregnation and calcination. The resulting materials were characterized with XPS, TEM, PXRD, EDX and Raman spectroscopy. Sodium hydroxide treatment led to the purification of MWCNTs (p-MWCNTs). Functionalizing purified MWCNTs with in situ generated tricarboxylic aryl diazonium salts (p-MWCNTs-D3) followed by impregnation using IR irradiation was the key step in efficiently impregnating p-MWCNTs-D3 with lanthanum nitrate (p-MWCNTs-D3/LaN). Calcination of p-MWCNTs-D3/LaN at 500 °C under argon atmosphere resulted in homogeneously distributed LaH NPs on MWCNTs (p-MWCNTs/LaH) in the size range of 1.5–6.0 nm, with a Gaussian mean diameter of 3.0 nm. LaH nanoparticles were found to be amorphous. This method is applicable to large-scale production, which opens interesting outlooks for nanotechnology applications such as energy storage material and sensor for detection of chemical warfare agents in the defense sector.

Keywords Multi-wall carbon nanotubes · Lanthanum nitrate · Diazonium chemistry · Infrared irradiation · Lanthanum hydroxide nanoparticles · Decoration

---

A. K. Bhakta (✉) · J. Delhalle · Z. Mekhalif
Laboratory of Chemistry and Electrochemistry of Surfaces, University of Namur, 61 Rue de Bruxelles, 5000 Namur, Belgium
e-mail: arvind.bhakta@unamur.be

S. Kumari · S. Hussain
Department of Chemistry, Indian Institute of Technology Patna, Bihta, Bihar 801106, India

R. J. Mascarenhas
Electrochemistry Research Group, Department of Chemistry, St. Joseph's College, Lalbagh Road, Bangalore, Karnataka 560027, India

P. Martis
Loyola Centre for Research and Innovation, St. Aloysius College, Light House Hill Road, Mangalore, Karnataka 575003, India

R. Fangueiro and S. Rana (eds.), *Advanced Materials for Defense*, 4, Springer Proceedings in Materials, https://doi.org/10.1007/978-3-030-34123-7_3

1 Introduction

Carbon nanotubes (CNTs) are an example of the allotropes of one-dimensional carbon. Since their discovery in 1991 by Sumio Iijima, as a by-product of fullerenes which was produced by an arc discharge process [1], they have received great attention due to their unique properties such as high chemical stability, electrical and thermal conductivities, tensile strength and large surface area [2]. These CNTs properties have potential applications in the fields of solar and fuel cells, hydrogen storage, lithium ion batteries, sensors and supercapacitors [3], to cite a few. They are also considered to be an efficient template for the assembly of nanoparticles [4–6].

CNTs are usually classified in two categories: single wall carbon nanotubes (SWCNTs) and multi-wall carbon nanotubes (MWCNTs). Most of as-synthesized CNTs contain different impurities (amorphous carbon, nanocrystalline graphite, fullerenes, metallic catalysts, silica and alumina) and are insoluble in many solvents [7]. Various methods have been developed to functionalize CNTs and to enhance their solubility and dispersion in different solvents. CNTs functionalization plays also a major role in tuning their properties [8].

Lanthanides are very interesting compounds having distinctive optical, magnetic and catalytic properties. This is due to their unique electronic configuration based on 4f electrons [9]. Lanthanum hydroxides are particularly important since the hydroxyl groups are active sites for surface functionalization through condensation reaction. It has been widely used in hydrogen storage materials, superconductive materials, high-potential oxide ceramics, supercapacitors [10] and adsorption of phosphate ions [11]. The catalytic and sorbet properties of lanthanum hydroxide have been recently considered for their potential applications [12, 13].

CNTs absorb near-infrared radiations (IR), promptly transferring electronic excitations into molecular vibration energies which results in heat. These very interesting photo-absorption and photo-thermal properties have been also used for CNTs decoration [14]. It has been proven that the heat generated by the molecular vibrations of CNTs (IR irradiation method) results in better hybrid formation in comparison with traditional heating [15].

In the present approach, we aim to control the size, concentration, nature and distribution of lanthanum hydroxide nanoparticles (LaH NPs) on MWCNTs using a simple and efficient method involving diazonium chemistry and IR irradiation. To the best of our knowledge there are no reports on LaH NPs decoration on tricarboxylic aryl diazonium functionalized MWCNTs using IR irradiation. The hybrid materials resulted from the combination of MWCNTs and LaH NPs are expected to lead to effective integration of properties of both constituents. This is important for nanotechnology applications such as in energy storage (lithium ion batteries, super capacitors and hydrogen storage) and conversion devices (solar cells and fuel cells), sensor for detection and removal of chemical warfare agents in defense sector.

2 Materials and Methods

2.1 *Raw Materials*

MWCNTs (NC 7000) (>95%) were received from Nanocyl SA (Belgium) with length in the range of 0.1–10 μm and an average diameter of 10 nm. Sodium nitrite, 99.2% (Fisher Scientific UK), 5-amino-1,2,3-benzenetricarboxylic acid, 97% (ABCR GmbH & CO. KG), perchloric acid, 70% (Merck Eurolab nv/sa), acetone, >99% (Chem-Lab), NaOH, ≥98% (ACS Reagent) and pentane, 99% (Lab-Scan Analytical Sciences), were used without further purification. Aqueous solutions were prepared using ultra-pure water.

2.2 *Purification and Functionalization of MWCNTs*

Purification and functionalization steps were performed as described in our previous work [16]. The obtained purified MWCNTs and tricarboxylic aryl diazonium functionalized MWCNTs were referred to as p-MWCNTs and p-MWCNTs-D3, respectively.

2.3 *Impregnation of Lanthanum Nitrate on Functionalized MWCNTs*

Tricarboxylic aryl diazonium functionalized MWCNTs (p-MWCNTs-D3) were added to a 100 ml of 0.01 M lanthanum nitrate aqueous solution and subjected to 5 min sonication (UP200S Ultrasonic Processor with cycle 1 and amplitude % 100) followed by 2 h IR irradiation under constant magnetic stirring. The mixture was cooled down to room temperature and filtered. The residue was washed with water (3 times), then with acetone (twice) and ultimately dried in air. The impregnated p-MWCNTs-D3 were referred to as p-MWCNTs-D3/LaN.

2.4 *Synthesis of p-MWCNTs/LaH*

MWCNTs-D3/LaN were calcined at 500 °C under a continuous flow of argon gas for 2 h. The resulting materials were labelled to as p-MWCNTs/LaH.

2.5 Characterization Techniques

X-ray photoelectron spectroscopy (XPS) spectra were recorded using a Thermo Scientific K-Alpha spectrometer with a monochromatized Al Kα radiation (1486.6 eV) for studying the chemical composition of the materials. Transmission electron microscopy (TEM) analyses were carried out using a Tecnai 10 philips microscope. It was operated at 80 kV accelerating voltage and 5μA emission current. Powder X-Ray diffraction (PXRD) analyses were carried out using PAN analytical XPert PRO Bragg-Brentano diffractometer. Energy dispersive X-ray (EDX) analyses were performed out at 15 kV accelerating voltage and 20 μA emission current. An IR lamp (capacity: 150 W, 50/60 Hz, voltage: 230 V) was used for the irradiation of the samples. Raman study was carried out using Princeton instruments Acton SP2750. Instrument was operated with excitation laser: Ar 514.5, exposure time: 5 s and number of accumulations: 10.

3 Results and Discussion

The materials were characterized using XPS, TEM, PXRD, EDX and Raman.

3.1 Materials Chemical Composition by XPS

Table 1 displays the atomic percentage compositions of the materials obtained from XPS analyses. The absence of alumina in p-MWCNTs (Table 1; Fig. 1a) proves the efficiency of the purification method. The increase of O and the presence of N in p-MWCNTs-D3 (Fig. 1b) support the fact that the p-MWCNTs were functionalized with tricarboxylic aryl diazonium. The presence of La in p-MWCNTs-D3/LaN (Fig. 1c) indicates the effective impregnation on p-MWCNTs-D3 with lanthanum. The presence of lanthanum was also evidenced in p-MWCNTs/LaH (Fig. 1d). The high resolution XPS spectrum of La3d region of the p-MWCNTs/LaH (Fig. 1e) gave $\Delta BE = 3.7$ eV, the difference between the binding energy of the doublet peak present

Table 1 Atomic percentages obtained from XPS analysis of the materials

Material	C%	O%	N%	La%	Al%
Crude MWCNTs	97.31	1.89	–	–	0.80
p-MWCNTs	98.2	1.80	–	–	–
p-MWCNTs-D3	72.91	23.89	3.20	–	–
p-MWCNTs-D3/LaN	82.85	14.01	1.34	1.80	–
p-MWCNTs/LaH	90.27	7.42	0.81	1.50	–

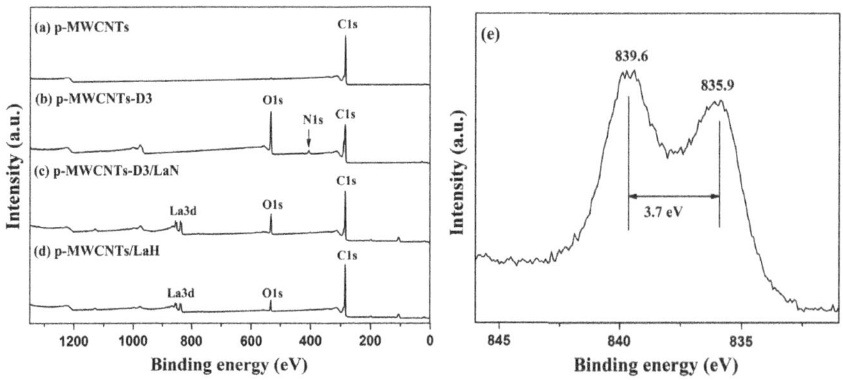

Fig. 1 XPS survey spectra of **a** p-MWCNTs, **b** p-MWCNTs-D3, **c** p-MWCNTs-D3/LaN, **d** p-MWCNTs/LaH and **e** high resolution XPS spectrum of La3d of p-MWCNTs/LaH

at 839.6 and 835.9 eV, revealed the presence of La mainly in its hydroxide form [17, 18]. This is further ascertained by the fact that the O1s signal appeared at 532.3 eV [17, 19] which has confirmed the presence of lanthanum in the hydroxide form. After the calcination process, a decrease in the atomic percentages of O, N and La was observed. It can be explained by the evaporation of some amount of lanthanum hydroxide and the breakage of few C–N diazonium connections due to the heating process.

The mechanism [20] through which diazonium salts $R\text{-}N_2^+ X^-$ (R = aryl in this work) were formed has been reported in the literature. In this process, diazonium ions reduction through electron transfer and dinitrogen loss either led to the formation of an aryl radical which reacted with the carbon nanotube surface to form a C–C covalent bond or the diazonium directly reacted with carbon atom of the nanotube to form C–N covalent bonds [21]. These two possibilities are shown in Fig. 2, where it also explains the overall process involved in the decoration of MWCNTs with lanthanum hydroxide nanoparticles.

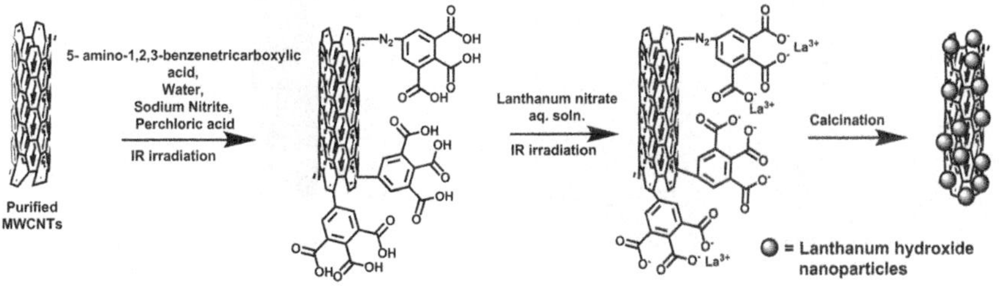

Fig. 2 Overall steps in the decoration MWCNTs with lanthanum hydroxide nanoparticles

3.2 Materials Morphology by TEM

Figure 3a shows the presence of a large alumina inclusion which was removed after the purification process (Fig. 3b) (representative figure). The treatment of crude MWCNTs with sodium hydroxide aqueous solution at 170 °C resulted in the conversion of alumina to water soluble sodium aluminate (Eq. 1) which was discarded after filtration and several washing steps with water.

$$Al_2O_3 + 2NaOH \rightarrow 2NaAlO_2 + H_2O \quad (1)$$

This alkaline treatment has advantages over an acid treatment which induces cutting of MWCNTs and the formation of undesired reaction products (e.g. OH, COOH etc.) on their surfaces [22, 23]. As evidenced from TEM analyses (Fig. 3c, d), very small size LaH NPs (size range: 1.5–6.0 nm and a Gaussian mean diameter of ~3.0 nm) are uniformly distributed on the MWCNTs and strongly anchored (no nanoparticles detached from the MWCNTs were observed).

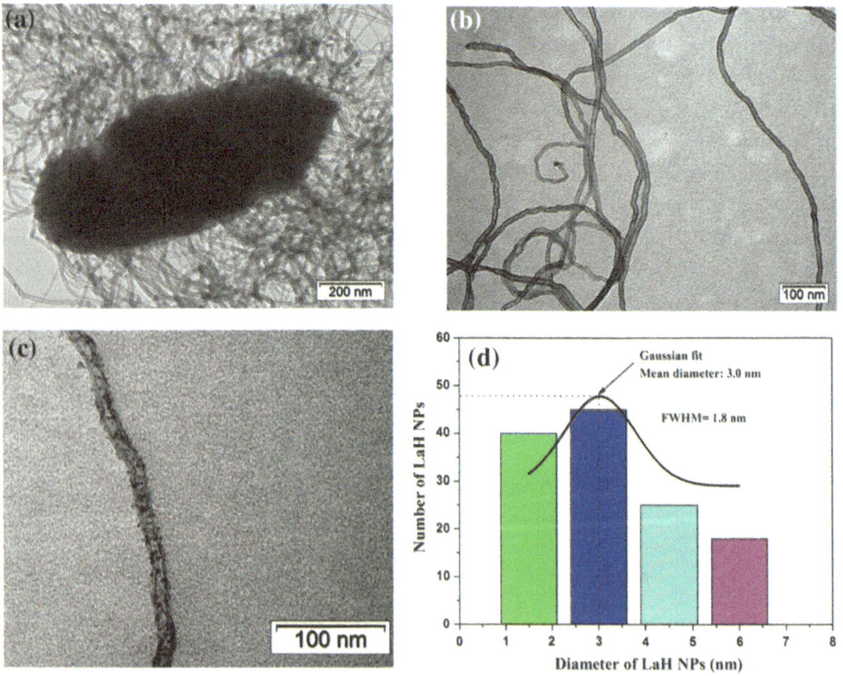

Fig. 3 TEM image of **a** crude MWCNTs, **b** p-MWCNTs and **c** p-MWCNTs/LaH. **d** Diameter distribution of LaH NPs on the p-MWCNTs/LaH surface

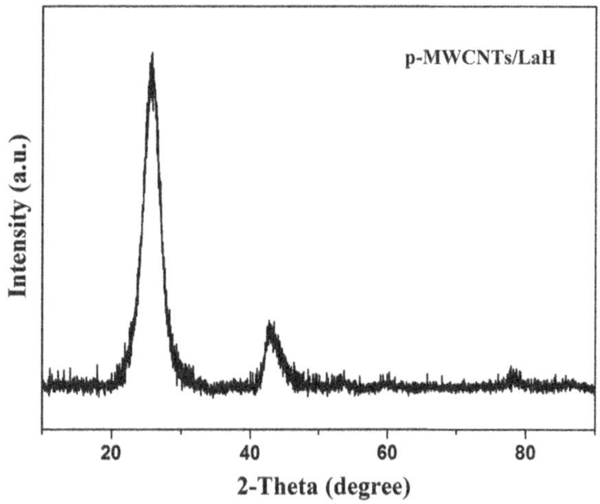

Fig. 4 PXRD pattern of p-MWCNTs/LaH

3.3 PXRD Characterisation of the Materials

The intense diffraction peak at $2\theta = 25.63°$ is due to the reflection from (002) plane of graphitic carbon in the MWCNTs structure [24]. Comparatively, the less intense peak at $2\theta = 42.81°$ is due to the reflection from (100) plane of MWCNTs. No signal corresponding to LaH NPs (Fig. 4) was observed indicating the amorphous nature of these particles.

3.4 EDX Mapping of the Materials

The EDX analysis confirmed that crude MWCNTs contained alumina impurities which were eliminated after purification (Table 2). The EDX data (Table 2) do not fit with XPS results (Table 1) for crude MWCNTs and p-MWCNTs/LaH, which is not surprising since these two techniques have different probing depths (XPS: ~10 nm and EDX: ~1 to 2 μm). Alumina was better detected in EDX compared to XPS due to its higher sensitivity led to the atomic % differences in the case of crude MWCNTs. For p-MWCNTs/LaH, comparatively less La in EDX data indicate that La is present on the surface of the MWCNTs. Because of the low amount of N and the overlap of

Table 2 Atomic percentages obtained from EDX analysis for the various materials

Materials	C%	O%	N%	Al%	La%
Crude MWCNTs	91.76	6.54	–	1.70	–
p-MWCNTs	98.31	1.65	–	0.04	–
p-MWCNTs/LaH	93.09	5.93	–	–	0.98

the N peak (0.392 keV) with those of C (0.277 keV) and O (0.525 keV), N was not identified in the p-MWCNTs/LaH EDX spectrum (not given).

3.5 Raman Characterization of Materials

The Raman spectra of crude, purified and lanthanum hydroxide nanoparticles decorated MWCNTs are presented in Fig. 5. The important characteristics in the Raman spectra are the D band which is found at around 1347 cm^{-1} and the G band appeared at around 1582 cm^{-1}. The parameter used to probe the quality of CNTs using Raman technique is the ratio between the intensities of the D and G bands (I_D/I_G). The I_D/I_G ratio in Raman spectrum of crude MWCNTs is larger than 1 indicating that crude MWCNTs possess inherent defects formed during their synthesis. The I_D/I_G ratio decreases after purification step meaning that the quality of CNTs increased after purification procedure. This result supports the importance of purification. This ratio further decreases in lanthanum hydroxide nanoparticles decorated MWCNTs indicating that the quality of MWCNTs was further enhanced which can be due to the thermal annealing of the defected MWCNTs. Also, this confirmed that the adopted procedure preserves the integrity of MWCNTs.

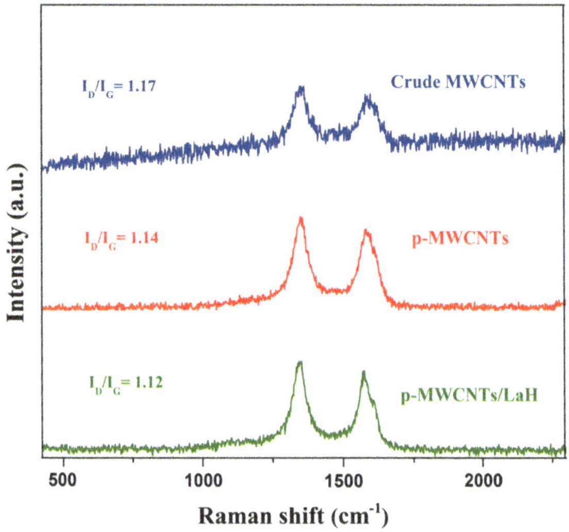

Fig. 5 Raman spectra of **a** crude MWCNTs, **b** purified MWCNTs and **c** p-MWCNTs/LaH

4 Conclusions

A simple and efficient technique has been developed to decorate multi-wall carbon nanotubes with lanthanum hydroxide nanoparticles. Amorphous lanthanum hydroxide nanoparticles in the size range of 1.5–6.0 nm, with a Gaussian mean diameter of ~3.0 nm, were homogeneously distributed on the MWCNTs surfaces. Applicability of this material to large-scale production opens interesting outlooks for different applications including those in the defense sector such as energy storage and generation material, sensor for detection and removal of chemical warfare agents.

Acknowledgements Arvind K. Bhakta thanks the University of Namur for a CERUNA doctoral fellowship.

Conflict of Interest The authors declare that there is no conflict of interest.

References

1. Iijima S (1991) Helical microtubules of graphitic carbon. Nature 354:56–58. https://doi.org/10.1038/354056a0
2. Ajayan PM (1999) Nanotubes from carbon. Chem Rev 99:1787–1799. https://doi.org/10.1021/cr970102g
3. Favero VO, Oliveira DA, Lutkenhaus JL, Siqueira JR Jr (2018) Layer-by-layer nanostructured supercapacitor electrodes consisting of ZnO nanoparticles and multi- walled carbon nanotubes. J Mater Sci 53:6719–6728. https://doi.org/10.1007/s10853-018-2010-4
4. Bhakta AK, Mascarenhas RJ, D'Souza OJ, Satpati AK, Detriche S, Mekhalif Z, Dalhalle J (2015) Iron nanoparticles decorated multi-wall carbon nanotubes modified carbon paste electrode as an electrochemical sensor for the simultaneous determination of uric acid in the presence of ascorbic acid, dopamine and l-tyrosine. Mater Sci Eng C 57:328–337. https://doi.org/10.1016/j.msec.2015.08.003
5. Bhakta AK, Detriche S, Kumari S, Hussain S, Martis P, Mascarenhas RJ, Delhalle J, Mekhalif Z (2018) Multi-wall carbon nanotubes decorated with bismuth oxide nanocrystals using infrared irradiation and diazonium chemistry. J Inorg Organomet Polym Mater. https://doi.org/10.1007/s10904-018-0800-4
6. Bhakta AK, Mascarenhas RJ, Martis P, Delhalle J, Mekhalif Z (2018) Multi-wall carbon nanotubes decorated with barium oxide nanoparticles. Synth Catal Open Access 3. https://doi.org/10.4172/2574-0431.100019
7. Karousis N, Tagmatarchis N, Tasis D (2010) Current progress on the chemical modification of carbon nanotubes. Chem Rev 110:5366–5397. https://doi.org/10.1021/cr100018g
8. Kaynak N, Aysen O, Karahasanog M (2018) Photoactive multi-walled carbon nanotubes: synthesis and utilization of benzoin functional MWCNTs. J Mater Sci 53:9598–9610. https://doi.org/10.1007/s10853-018-2233-4
9. Salavati-niasari M, Hosseinzadeh G, Davar F (2011) Synthesis of lanthanum hydroxide and lanthanum oxide nanoparticles by sonochemical method. J Alloys Compd 509:4098–4103. https://doi.org/10.1016/j.jallcom.2010.07.083
10. Ho K, Liu H, Ke QQ, Mao L, Hu Y, Xu L, Wang J (2017) Nickel and lanthanum hydroxide nanocomposites with much improved electrochemical performance for supercapacitors. J Am Ceram Soc 100:247–256. https://doi.org/10.1111/jace.14546

11. Zong E, Liu X, Wang J, Yang S, Jiang J, Fu S (2017) Facile preparation and characterization of lanthanum- loaded carboxylated multi-walled carbon nanotubes and their application for the adsorption of phosphate ions. J Mater Sci 52:7294–7310. https://doi.org/10.1007/s10853-017-0966-0
12. Ghiasi M, Malekzadeh A (2015) Synthesis, characterization and photocatalytic properties of lanthanum oxy-carbonate, lanthanum oxide and lanthanum hydroxide nanoparticles. Superlattices Microstruct 77:295–304. https://doi.org/10.1016/j.spmi.2014.09.027
13. Ziaee F, Gholizadeh M, Seyedi SM (2018) Magnetically recoverable lanthanum hydroxide as an efficient catalyst for aerobic oxidative conversions of primary alcohols to the nitriles. Appl Organometal Chem 1–8. https://doi.org/10.1002/aoc.4253
14. Martis P, Venugopal BR, Seffer J-F, Delhalle J, Mekhalif Z (2011) Infrared irradiation controlled decoration of multiwalled carbon nanotubes with copper/copper oxide nanocrystals. Acta Mater 59:5040–5047. https://doi.org/10.1016/j.actamat.2011.04.061
15. Venugopal BR, Detriche S, Delhalle J, Mekhalif Z (2012) Effect of infrared irradiation on immobilization of ZnO nanocrystals on multiwalled carbon nanotubes. J. Nanopart Res 14:1079. https://doi.org/10.1007/s11051-012-1079-y
16. Bhakta AK, Detriche S, Martis P, Mascarenhas RJ, Delhalle J, Mekhalif Z (2017) Decoration of tricarboxylic and monocarboxylic aryl diazonium functionalized multi-wall carbon nanotubes with iron nanoparticles. J Mater Sci 52:9648–9660. https://doi.org/10.1007/s10853-017-1100-z
17. Sunding MF, Hadidi K, Diplas S, Løvvik OM, Norby TE, Gunnæs AE (2011) XPS characterisation of in situ treated lanthanum oxide and hydroxide using tailored charge referencing and peak fitting procedures. J Electron Spectrosc Relat Phenomena 184:399–409. https://doi.org/10.1016/j.elspec.2011.04.002
18. Bosch CE, Copley MP, Eralp T, Bilbé E, Thybaut JW, Marin GB, Collier P (2017) Tailoring the physical and catalytic properties of lanthanum oxycarbonate nanoparticles. Appl Catal A Gen 536:104–112. https://doi.org/10.1016/j.apcata.2017.01.019
19. Lacombe S, Geantet C, Mirodatos C (1994) Oxidative coupling of methane over lanthana catalysts. J Catal 151:439–452
20. Chehimi MM (ed) (2012) Aryl diazonium salts: new coupling agents in polymer and surface science. Wiley-VCH Weinheim Germany. ISBN 978-3-527-32998-4
21. Bhakta AK, Kumari S, Hussain S, Martis P, Mascarenhas RJ, Delhalle J, Mekhalif Z (2019) Synthesis and characterization of maghemite nanocrystals decorated multi-wall carbon nanotubes for methylene blue dye removal. J Mater Sci 54:200–216. https://doi.org/10.1007/s10853-018-2818-y
22. Kilinç E (2016) γ-Fe_2O_3 magnetic nanoparticle functionalized with carboxylated multi walled carbon nanotube: Synthesis, characterization, analytical and biomedical application. J Magn Magn Mater 401:949–955. https://doi.org/10.1016/j.jmmm.2015.11.003
23. Chen J, Hamon MA, Hu H, Chen Y, Rao AM, Eklund PC, Haddon RC (1998) Solution properties of single-walled carbon nanotubes. Science (80) 282:95–98. https://doi.org/10.1126/science.282.5386.95
24. Yan F, Liu L, Li M, Zhang M, Xiao L, Ao Y (2018) Preparation of carbon nanotube/ copper/ carbon fiber hierarchical composites by electrophoretic deposition for enhanced thermal conductivity and interfacial properties. J Mater Sci 53:8108–8119. https://doi.org/10.1007/s10853-018-2115-9

Ballistic Composites Reinforced with Inorganic Nanotubes of Tungsten Disulfide

Danica M. Simić, Dušica B. Stojanović, Novica Ristović, Milorad Zrilić, Zijah Burzić, Milica Marjanović, Petar S. Usković and Radoslav Aleksić

Abstract This research considers a possibility of improving the impact and ballistic resistance of composite materials based on aramid fabric impregnated with poly (vinyl butyral), PVB, with the application of multi-layer inorganic nanotubes of tungsten disulfide (INT-WS_2) as the reinforcement. Multi-layered nano-structures of tungsten disulfide are known by their excellent mechanical resistance, especially shock absorbing properties, which promote their use as a reinforcement in composite materials. INT-WS_2 used in this research were observed using scanning electron microscope and the obtained images were analyzed in program Image Pro Plus. Before the application in the composite they were ultrasonically deagglomerated. Ballistic composites were prepared impregnating the aramid fabric with system PVB/INT-WS_2 and by pressing defined number of layers under high pressure and defined temperature. Charpy impact test, ballistic shooting with two different ammunition calibers and stab resistance test with controlled knife penetration were performed on prepared composite samples, and resulted in significant improvements for samples with nano-reinforcement regarding the absorbed energy of impact, toughness and the ballistic resistance.

Keywords Ballistic resistance · Tungsten disulfide nanotubes · Composites · Impact energy · Toughness

1 Introduction

Composite materials used in ballistic protection are most often based on fibers or a fabric of high toughness and tensile strength, such as aramid fabric, impregnated with some thermoplastic or thermosetting polymer. Ultra-high molecular weight polyethylene (UHMWPE) is also widely used [1, 2]. There is also a tendency to achieve additional reinforcement of ballistic composite systems by introducing nanoparticles

D. M. Simić (✉) · N. Ristović · Z. Burzić · M. Marjanović
Military Technical Institute (MTI, VTI), Belgrade, Serbia
e-mail: simic_danica@yahoo.com

D. B. Stojanović · M. Zrilić · P. S. Usković · R. Aleksić
Faculty of Technology and Metallurgy, University of Belgrade, Belgrade, Serbia

R. Fangueiro and S. Rana (eds.), *Advanced Materials for Defense*, 4, Springer Proceedings in Materials, https://doi.org/10.1007/978-3-030-34123-7_4

of different properties and structures [3–8]. In this research a possibility is investigated of making hybrid nanocomposite materials of improved impact resistance, applying inorganic multi-layer nanotubes of tungsten disulfide (INT-WS_2) as the reinforcement in poly (vinyl butyral), PVB, as impregnation on aramid fabric.

Poly (vinyl butyral), a polymer known by its good adhesion, good tensile strength and elasticity, is widely used in coatings, binders, as well as a polymer matrix in composite materials for protective helmets and ballistic vests [9–12]. In these composites, PVB is actually combined with phenolic resin. These two resins actually react, which prevents the brittle fracture and enhances the impact resistance of phenolic resin [10].

Tungsten disulfide, as most of the transition metal chalcogenides, has layered structure: two layers of close-packed chalcogenide atoms with one metal layer between them, held together with van der Waals forces. Nanoparticles of such inorganic compounds with layered structures are unstable in the plane, 2D form, and they tend to fold on themselves forming 3D particles: seamless hollow structures like multiwall nanotubes and fullerene-like spherical nanoparticles, with extraordinary properties [13–17]. These multi-layered nano-structures of tungsten disulfide are known by their high pressure, impact and heat resistances; high shock absorbing properties, thermal stability, chemical inertness, compatibility with various systems, and all these advantages encourage their use as a promising reinforcement of polymers, coatings and composites [18–28]. In earlier researches it has been proven that nanostructures of WS_2 successfully reinforce PVB [29–31].

Within this research laminated composite structures were manufactured in a simple technology as most of ballistic prepregs are made: applying an impregnating solution of polymer on the aramid fabric [11, 12]. After drying and solvent release, the layers of aramid fabric were pressed.

2 Experiments and Methods

2.1 WS_2 Nanotubes Characterization

Before their application in composite, nanotubes of WS_2 (as received) were observed using scanning electron microscope JEOL JSM-6610 LV with lantan hexaboride cathode. The nanotubes were coated with gold and observed under 20 kV of voltage. A chosen representative image of deagglomerated nanotubes was analyzed in program Image Pro Plus in order to determine actual diameters of WS_2 nanotubes.

2.2 Ballistic Composites Manufacture

For this research, composite samples of laminated multi-layer structures were prepared, consisting of the following materials:

- aramid fabric (aramid prepreg with a thin layer of phenolic resin: Kevlar Prepreg 258HPP, DuPont)
- PVB powder, fine grained (Mowital B60H, Kuraray GmbH)
- INT-WS_2 nanotubes of average diameter 80–100 nm and length 10–20 μm (NanoMaterials Ltd., Israel),
- ethanol (Zorka Pharma, Serbia).

Properties of the INT-WS_2, provided by the supplier, are declared as follows:

- Purity: >99%,
- Particle density, at 25 °C: 7 g/cm^3,
- Apparent (bulk) density: 1.7–3.5 g/cm^3,
- Nanotube length: 10–20 μm,
- Oxidation stability, in air: >350 °C, in inert: >1000 °C,
- Molecular weight: 247.98 g/mol,
- Shock wave resistance: >21 GPa,
- Maximum elongation: 14%,
- Young's module of elasticity: 170 GPa,
- Solubility: insoluble in oil, grease, silicone fluids, alcohols and fuels.

Tungsten disulfide nanotubes were first deagglomerated and dispersed by an ultrasonic processor (Badelin SonoPuls) in ethanol, and PVB powder was added to make 10 wt% solution of PVB in it. This solution was applied on aramid fabric. In this system, a small amount of INT-WS_2 was used, 3 wt% regarding the mass of PVB, i.e. 0.3 wt% regarding the total mass of composite material. After ethanol complete evaporation, 8 or 20 pieces of fabric impregnated with the composition PVB/INT-WS_2 were pressed on a hydraulic press (Belišće) in order to form laminated composite samples for tests. Processing parameters were defined: temperature approximately 165 °C (optimal temperature for curing reaction between PVB and phenolic resin), maximum pressure of 100 bars and total duration of pressing 24 min (4 min with contact pressure with 4 time opening to release the evolved gases, and 20 min at maximum pressure). This procedure was also used for the manufacture of the composite without INT-WS_2 reinforcement.

This way the samples were prepared for mechanical testing of impact resistance and for ballistic tests. Samples with 8 layers were intended for Charpy impact test and stab resistance test, while samples with 20 layers were intended for ballistic shooting test.

2.3 *Impact and Ballistic Resistance Tests*

Impact strength was examined using Charpy impact pendulum (Zwick). Unnotched specimens were used, with dimensions approximately 60 mm × 10 mm × 3 mm, that were cut out from pressed composites using an underrated bandsaw.

The ballistic resistance of the composites was examined trough stab resistance test and ballistic shooting test. The stab resistance tests were carried out according to the standard NIJ 0115.00, at defined knife speed, 100 mm/min, and with defined knife maximum movement, i.e. displacement, 50 mm. Each sample was stabbed three times. Ballistic shooting tests were carried out in accordance with the standard NIJ 0101.06, using two different ammunition calibers, full metal jacketed FMJ 9 mm and 357 Magnum. The sample dimensions for shooting tests were 300 mm × 250 mm, with 20 pressed layers. Tests were performed with three shots of both calibers. Backface deformation, BFD, imprinted in the ballistic clay placed behind the tested samples, was measured after each shot.

3 Results and Discussions

3.1 Nanotubes SEM Analysis

SEM images of INT-WS_2 are given in Fig. 1. It may be observed that INT-WS_2, as received, are highly agglomerated, so it is necessary, for samples preparation, to apply ultrasound to deagglomerate the nanotubes before their incorporation in composite material.

Image of deagglomerated nanotubes was analyzed in program Image Pro Plus, and for a sample of 50 measurements, nanotubes diameter distribution was obtained (Fig. 2).

It can be noted that nearly 70% of the nanotubes have diameters in the range 80–140 nm.

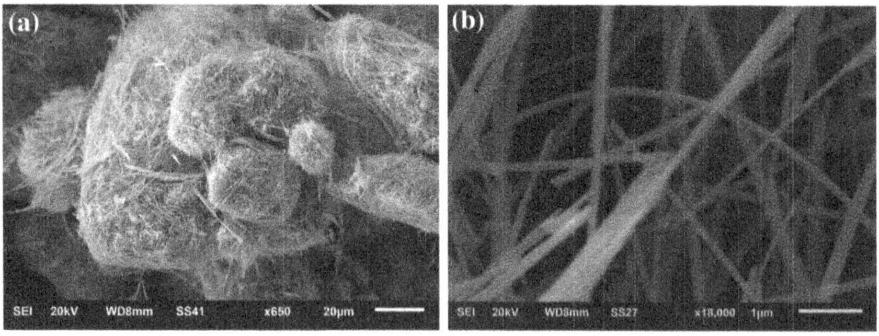

Fig. 1 SEM images of agglomerated **a** and deagglomerated INT-WS_2 **b**

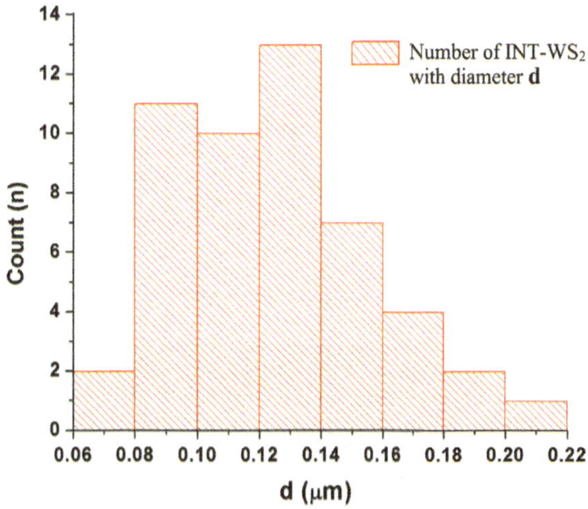

Fig. 2 Nanotubes diameter distribution obtained from Image Pro Plus analysis

Table 1 Charpy impact test results

Sample	Sample width [mm]	Sample thickness [mm]	Absorbed impact energy, E [J]	Impact toughness, U_T [J/cm^2]
Aramid/PVB	9.56	3.17	4.60	15.18
Aramid/PVB/INT-WS_2	9.09	3.10	5.96	21.15

3.2 *Impact Resistance*

The results from the Charpy impact test (average values of absorbed impact energy and impact toughness) are given in Table 1. Impact toughness is defined as the absorbed energy per unit of the cross-section area of the sample.

An evident improvement of impact resistance has been observed for the composite reinforced with INT-WS_2: absorbed energy increase for 29.6% and an increase of impact toughness for 39.3%.

3.3 *Ballistic Resistance*

Within the test of resistance to knife stab, the results were observed as follows: depth of deformation in the ballistic clay was measured, as in Fig. 3 [32], where H is the measured depth after knife pull-out, P is the perforation depth in the clay. However, since the total trauma depth is determined by the total knife movement, in this case 50 mm = H + P, the depth P is not relevant. The depth H was measured and analyzed

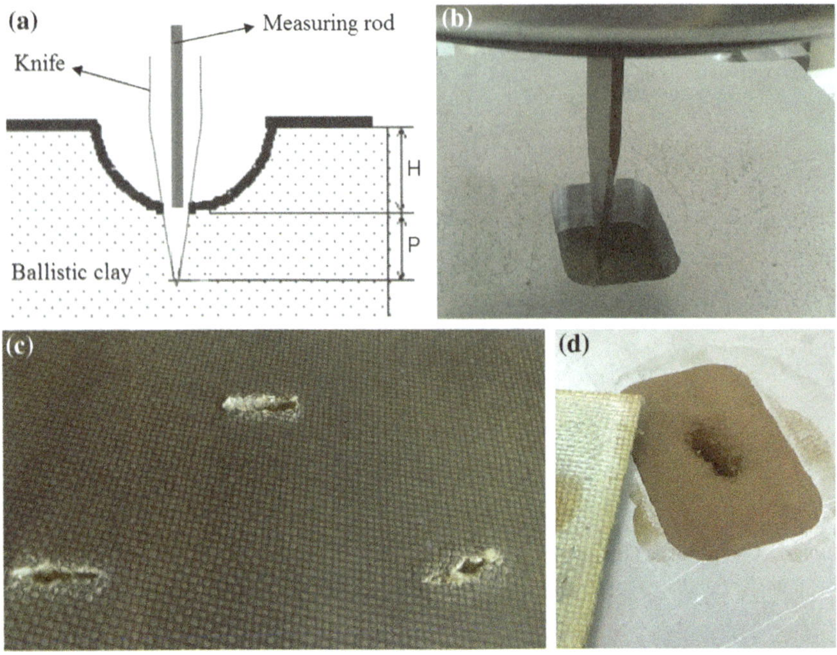

Fig. 3 Scheme of measuring the depth of the penetration/material deformation after the knife stab **a**, testing stab resistance **b**, sample aramid/PVB/INT-WS_2 after knife penetrations **c** and a puncture tray in ballistic clay **d**

as a deformation of the material caused by the knife stab, until the point when the puncture happened.

The stab resistance of the composites is quantified through an average value of the deformation depth, of the maximum registered force (F_{max}), and the absorbed energy of the stab (Table 2). The absorbed energy of the stab is determined as the calculated area between the force-deformation curve and x-axis, integral of the function force-displacement, given in diagrams in Fig. 4. This value represents the measure of the toughness of the material [33, 34].

The absorbed energy of the stab is 8.5% higher for the composite material with INT-WS_2, but even more important, deformation depth for the samples containing INT-WS_2 is for 40.4% smaller than for the sample without the nanotubes.

The results of the ballistic shooting tests are shown in Table 3, showing measured back face deformations after bullet impact.

Table 2 Stab resistance test results, 8-layered samples

Sample	Deformation depth [mm]	F_{max} [N]	Absorbed energy, E [J]
Aramid/PVB	4.63	1820	69,890
Aramid/PVB/INT-WS_2	2.76	1910	75,858

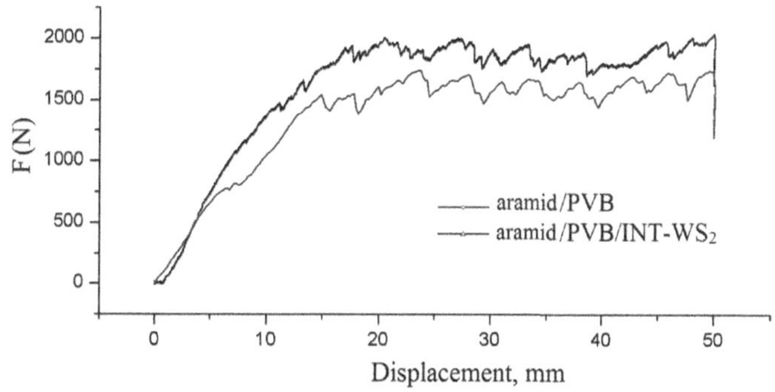

Fig. 4 Force-displacement diagrams obtained from stab resistance testing

Table 3 Ballistic test results, 20-layered samples

Sample/shooting		9 mm FMJ			0.357 Magnum		
		Bullet velocity [m/s]	BFD [mm]		Bullet velocity [m/s]	BFD [mm]	
Aramid/PVB	1	387	16	15.5	441	14	
	2	387	15		430	Penetration	
	3	391	15.5		–	–	
Aramid/PVB/INT-WS_2	1	382	15	13.8	438	16	14.3
	2	387	13		438	16	
	3	393	13.5		439	11	

This test has once again confirmed the reinforcing effect of INT-WS_2, resulting in 12% smaller BFD depths in composite with the nanotubes and in successful stopping the shots of both calibers, unlike the sample without nanoreinforcement. INT-WS_2, as well as fullerene-like nanostructures of WS_2, have a notable positive impact on mechanical resistance of composites for advanced applications [35]. In further research it would be interesting to apply different concentrations of INT-WS_2, as well as to examine samples with different number of aramid fabric layers.

4 Conclusions

Composite material based on aramid fabric and PVB has been reinforced by adding small concentration of INT-WS_2. Significant improvement of anti-ballistic performance has been observed for the samples of these new laminated composite structures containing the nanoreinforcement: 39.3% higher impact toughness, 8.5% higher absorbed energy of stab, and 12% smaller depth of backface deformation after shots.

The presented results for new composite material based on aramid fabric impregnated with PVB reinforced with INT-WS_2 affirm their possible application in ballistic protection equipment and as well in other areas, like automotive and aircraft industry.

Acknowledgements The authors gratefully acknowledge the funding by Serbian Ministry of Science, Technological Development and Education, under grants/projects TR 34034 and TR 34011.

References

1. Utracki LA (2010) Rigid ballistic composites. NRC Publications Archive, Canada
2. Iannucci L, Pope D (2011) High velocity impact and armour design. Express Polym Lett 5:262–272
3. Obradović V, Stojanović DB, Zivković I, Radojević V, Uskoković PS, Aleksić R (2015) Dynamic mechanical and impact properties of composites reinforced with carbon nanotubes. Fibers Polym 16(1):138–145
4. Obradović V, Stojanović DB, Jančić-Heinemann R, Zivković I, Radojević V, Uskoković PS, Aleksić R (2014) Ballistic properties of hybrid thermoplastic composites with silica nanoparticles. J Eng Fibers Fabr 9(4):97–107
5. Obradović V, Stojanović D, Kojović A, Zivković I, Radojević V, Uskoković P, Aleksić R (2014) Aramid composites impregnated with different reinforcement: nanofibers, nanoparticles and nanotubes. Zaštita Mater (Mater Prot) 15(4):351–361
6. Simić D, Stojanović D, Brzić S, Totovski L, Uskoković P, Aleksić R (2017) Aramid hybrid composite laminates reinforced with inorganic fullerene-like tungsten disulfide nanoparticles. Compos: Part B: Eng 123:10–18. https://doi.org/10.1016/j.compositesb.2017.05.002
7. Simić D, Stojanović D, Tasić A, Uskoković P, Aleksić R (2016) Effect of IF-WS_2 nanoparticles addition on thermo-rheological and mechanical behavior of aramid/phenolic resin/PVB composite material. In: Book of abstracts of The eighteenth annual conference YUCOMAT 2016, Herceg Novi, Montenegro, 5–10 Sept 2016, p 89
8. Ma T, Zhang T, Gao PG, Zhang JC (2013) Synthesis and properties of ultrahigh molecular weight polyethylene/WS_2 nanoparticle fiber for bullet-proof materials. Chin Sci Bull 58:945–948
9. http://www.kuraray.eu/en/produkte/productgroups/polyvinyl-butyral
10. Ismail IN, Ishak ZAM, Jaafar MF, Omar S, Zainal Abidin MF, Ahmad Marzuki HF (2009) Thermomechanical properties of toughened phenolic resol resin. Solid State Sci Technol 17(1):155–165
11. Dimeski D, Srebrenkoska V (2014) Advanced prepreg ballistic composites for military helmets. In: Proceedings of 6th international scientific conference on defensive technologies OTEH 2014, Belgrade, 09–10 Oct 2014, pp 418–424
12. Bhattacharyya D (1997) Composite sheet forming. Composite materials series, vol 11, Elsevier. ISBN 0-444-82641-6
13. Tenne R, Redlich M (2010) Recent progress in the research of inorganic fullerene-like nanoparticles and inorganic nanotubes. Chem Soc Rev 39:1423–1434
14. Zink N, Pansiot J, Kieffer J et al (2007) Chem Mater 19:6391–6400
15. http://nisusacorp.com/technology
16. Tenne R (2006) Inorganic nanotubes and fullerene-like nanoparticles. Nat Nanotechnol 1:103
17. Tenne R, Margulis L, Genut M, Hodes G (1992) Nature 360:444–446
18. Tevet O (2011) Mechanical and tribological properties of inorganic fullerene-like (IF) nanoparticles. Weizmann Institute of science, Rehovot, Israel
19. Kaplan-Ashiri I, Tenne R (2007) Mechanical properties of WS_2 nanotubes. J Cluster Sci 18:549–563

20. Xu F (2013) Large scale manufacturing of IF-WS_2 nanomaterials and their application in polymer nanocomposites. University of Exeter, Devon, UK
21. Zhu Y, Sekine T, Brigatti K, Firth S, Tenne R, Rosentsveig R, Kroto H, Walton D (2003) Shock-wave resistance of WS_2 nanotubes. J Am Chem Soc 125:1329–1333
22. Zhu Y, Sekine T, Li Y, Wang W, Fay M, Edwards H, Brown P, Fleischer N, Tenne R (2005) WS_2 and MoS_2 inorganic fullerenes—super shock absorbers at very high pressures. Adv Mater 17:1500–1503
23. Xu F, Kobayashi T, Yang Z et al (2017) ACS Nano 11:8114–8121
24. Volkova EI, Jones IA, Brooks R, Zhu Y, Bichoutskaia E (2013) Meso-scale modelling of shock wave propagation in a SiC/Al nanocomposite reinforced with WS_2-inorganic fullerene nanoparticles. Compos Struct 96:601–605
25. Simić D, Stojanović D, Kojović A, Dimić M, Totovski L, Uskoković P, Aleksić R (2016) Inorganic fullerene-like IF-WS_2/PVB nanocomposites of improved thermo-mechanical and tribological properties. Mater Chem Phys 184:335–344. https://doi.org/10.1016/j.matchemphys.2016.09.060
26. Lazić D, Simić D, Samolov A (2016) Effect of IF-WS_2 nanoparticles addition on physical-mechanical and rheological properties and on chemical resistance of water-based paints. In: Book of abstracts of the eighteenth annual conference YUCOMAT 2016, Herceg Novi, Montenegro, 5–10 Sept 2016, p 87
27. Lazić D, Simić D, Samolov A, Jovanović Desimir (2017) Properties of standard polymeric and water-based coatings for military camouflage protection with addition of inorganic fulerene-like tungsten disulphide (IF-WS_2) nanoparticles. Sci Tech Rev 67:38–44
28. Lazić D, Simić D, Samolov A (2016) Effect of IF-WS_2 nanoparticles addition on physical-mechanical and rheological properties and on chemical resistance of polymeric coating and polyurethane paint. In Proceedings of 7th international scientific conference on defensive technologies OTEH 2016, Belgrade, 6–7 Oct 2016, pp 609–613. ISBN 978-86-81123-82-9
29. Simić D, Stojanović D, Kojović A, Totovski L, Burzić Z, Uskoković P, Aleksić R (2014) Ultrasonic deagglomeration of tungsten disulfide nanoparticles (WS_2) in different solvents for enhancing nanomechanical properties in poly (vinyl butyral) (PVB) nanocomposites. In: Book of abstracts of the sixteenth annual conference YUCOMAT 2014, Herceg Novi, Montenegro, 1–5 Sept 2014, p 101
30. Simić D, Stojanović D, Kojović A, Dimić M, Uskoković P, Aleksić R (2017) Tribomechanical and thermal properties of PVB thin films reinforced with nano IF-WS_2. In: Book of abstracts COST MP1402 scientific workshop "ALD and related ultra-thin film processes for advanced devices", Belgrade, 29–30 Aug 2017, p 28
31. Simić D, Stojanović D, Dimić M, Totovski L, Brzić S, Uskoković, Aleksić R (2016) Preliminary analysis of the possibility of preparing PVB/IF-WS_2 composites. Effect of nanoparticles addition on thermal and rheological behavior of PVB. In: Proceedings of 7th international scientific conference on defensive technologies OTEH 2016, Belgrade, 6–7 Oct 2016, pp 618–623
32. Alpyildiz Tuba, Rochery Maryline, Kurbak Arif, Flambard Xavier (2011) Textile Stab and cut resistance of knitted structures: a comparative study. Res J 81(2):205–214. https://doi.org/10.1177/0040517510383617
33. Obradović V, Stojanović D, Jokić B, Zrilić M, Radojević V, Uskoković P, Aleksić R (2017) Nanomechanical and anti-stabbing properties of Kolon fabric composites reinforced with hybrid nanoparticles. Compos B Eng 108:143–152. https://doi.org/10.1016/j.compositesb.2016.09.095
34. Stojanović D, Zrilić M, Jančić-Heinemann R, Zivković I, Kojović A, Uskoković P, Aleksić R (2013) Mechanical and anti-stabbing properties of modified thermoplastic polymers impregnated multiaxial p-aramid fabrics. Polym Adv Technol 24:772–776. https://doi.org/10.1002/pat.3141
35. Simić D (2017) Ballistic hybrid nanocomposite materials reinforced with inorganic fullerenes. Doctoral dissertation, University of Belgrade, Faculty of Technology and Metallurgy, Belgrade, p vii

Graphene-Based Nanocomposites with Improved Mechanical and Ballistic Protection Properties

Israel Gago, Manuel del Río, Julián Carretero, Gerardo León, Isidro Ibarra and Beatriz Miguel

Abstract Body armour systems need to be improved for defeating new threats of asymmetric warfare and terrorism. Nowadays, new technologies like graphene-based nanocomposites constitute one of the most active research fields in high performance materials, due to the fact that this nanomaterial improves the properties of diverse polymers in a notable way. However, in the field of the application of these nanocomposites to improve the dynamic strength against supersonic impacts, experimental data are practically non-existent. Looking for improvements in this field, our research group has fabricated, following a similar procedure to that described in one of our previous patents, several laminated plates of graphene-based nanocomposites, composed of a polyester resin matrix doped with pristine few-layer graphene (FLG) and reinforced with a fibreglass woven fabric (FGRP), using doping percentages ranging between 0.25 and 1% in weight. The ballistic limit (V_0) of such plates against 7.62×51 mm NATO FMJ ammunition, fired with an Accuracy International AW sniper rifle, has been investigated in accordance with the NATO STANAG 2920 standard. The tensile mechanical properties (tensile strength and Young's modulus) and impact properties (Charpy impact strength) of the laminates were also characterised in accordance with UNE-EN ISO 527-4 and UNE-EN ISO 179-1 standards, respectively. The V_0 increases with the increment of the graphene doping percentage, reaching a maximum value of $266.4\,\mathrm{m\,s^{-1}}$ at 1% w/w, which means an improvement of a 72.2% with regard to the undoped FRGP laminate. The graphene doped samples also showed better tensile and impact properties. These results support the viability of the development of new graphene-based nanocomposites with improved mechanical and ballistic protection properties for security and defence applications.

I. Gago (✉) · M. del Río · J. Carretero · G. León · I. Ibarra · B. Miguel
Departamento de Ingeniería Química y Ambiental, Universidad Politécnica de Cartagena (UPCT), Paseo Alfonso XIII, 52, Cartagena, Spain
e-mail: israel.gago@upct.es

R. Fangueiro and S. Rana (eds.), *Advanced Materials for Defense*, 4, Springer Proceedings in Materials, https://doi.org/10.1007/978-3-030-34123-7_5

1 Introduction

Graphene was first isolated by Novoselov et al. in 2004 [1]. Since this achievement, it is generating an interest that has been growing every day in various technological sectors, both in the field of new materials in general and, more especially, in nanotechnology. Structurally, graphene is a sheet of monatomic thickness, formed by carbon atoms with sp^2 hybridization, linked together and forming an hexagonal crystalline lattice, in which the length of the C–C bond is 1.42 Å [2]. Although in addition to graphene there are other nanomaterials that have already been proposed for their application in the field of nanocomposites [3], recently it is this one which is generating more interest. This is because, although graphene stands out above all for the properties that postulate it to be the quintessential material in electronics and computer science of the future, it also has outstanding mechanical properties [4].

A personal ballistic protection system (PBPS) is an individual equipment whose functions are to prevent the penetration into the body of projectiles of firearms, fragments of shrapnel and/or bladed weapons, and to absorb the impact energy, reducing the Behind Armour Blunt Trauma (BABT) caused by it. Conventional vests are relatively flexible and light-weight because they are made of synthetic fibre fabrics arranged in successive layers, but this kind of systems only offer protection against impacts of handgun projectiles with deformable core, shotgun pellets and small fragments such as those of hand grenades. The protection against high-speed rifle bullets, larger shrapnel or very sharp objects, requires the use of rigid armour plates that normally overlap a conventional vest and have the great disadvantage of being heavy and uncomfortable, so that their use significantly limits the wearer mobility.

Since the discovery of Kevlar® [5] by Kwolek in 1971 [6], this polymer has been the material of choice for the manufacture of PBPS's for its excellent properties but, in recent years, new materials and fibres have been developed such as Twaron® [7], Dyneema® [8], Spectra® [9] or Zylon® [10]. Nevertheless, FGRP composites continue having a very extended use in all kinds of anti-ballistic systems due to their excellent protection/price ratio [11].

In the field of materials science, active research is being carried out into the development of second generation "*superfibres*", but these advances are complex, require large investments, and present many technical challenges still unresolved. Some recent research lines are aimed to integrate various nanomaterials in PBPSs. The first one, developed by researchers at the University of Delaware [12], is based on the infusion of silica nanoparticles (SiO_2) on the fibres surface, which causes in them a sufficient increase in stiffness to protect the wearer. These coatings, which have been defined as "*Shear Thickening Fluids*" (STF), have attracted the interest of several companies in the defence sector. However, currently, there is not any certified system based on this technology which is commercially available.

Some NATO armies have expressed interest in a new fibre woven from carbon nanotubes that has been developed by scientists at the University of Michigan, which has some mechanical properties that allow to assume its potential usefulness in this field [13]. There are also several studies about nanocomposites that integrate multilayer carbon nanotubes in epoxy matrices combined with ceramics [14]. However, all these technologies are still at very early stages of development.

At the end of 2014, a joint research team from Rice University and the Massachusetts Institute of Technology published a study on the mechanical properties of graphene against the impact of SiO_2 microspheres at high speed [15]. However, to date, it has not been published any study on the capabilities of graphene-based nanocomposites in terms of ballistic protection on a macroscopic scale.

In order to evaluate the feasibility of the development of new materials based on graphene for its possible application in ballistic protection systems, our research group has carried out this study about the influence that the addition of graphene to a composite material has on its tensile and impact mechanical properties, as well as its ballistic limit. This paper presents the first results obtained using a series of nanocomposites prepared from a polyester resin reinforced with fibreglass and doped with FLG ($\leq$7 layers).

2 Experimental

2.1 Pristine Few-Layer Graphene

FLG was synthesized in our laboratory and characterized by Raman spectroscopy (Jovin-Ivon LabRam, SiO_2/Si, 532 nm), high-resolution transmission electron microscopy or HR-TEM (JEOL JEM-2100, 200 kV) and X-ray fluorescence elemental analysis or XRF (BRUKER S4 PIONEER, multires-He_{36}). The final product obtained is shown in Fig. 1.

Fig. 1 FLG nanopowder produced by our research group (left) and image of the Tindall's effect obtained in a stable dispersion (distilled water) by a visible green laser of 512 nm (right)

2.2 Fabrication of Nanocomposites Laminated Plates

The materials used to manufacture the nanocomposites were an orthophthalic unsaturated polyester resin as matrix (Recapoli 955 INF DCPD), a bidirectional E fibreglass as reinforcing woven fabric (plain, 0°/90°, 600 g m^{-2}), and the FLG previously produced.

The preparation of the matrices of the doped nanocomposites was carried out in a similar way to that described in our patent with some modifications [16], using four different degrees of doping (0.25, 0.50, 0.75 and 1.00%) in weight. The manufacture of the laminates was carried out by a combined technique of hand lay-up and vacuum assisted molding under compression at room temperature. After the curing processes were completed, the specimens were machined in the form of square plates of 25 cm × 25 cm.

2.3 Mechanical Tests

For the characterization of tensile and impact mechanical properties of the manufactured laminates, corresponding series of specimens were machined according to the specifications described in the standards UNE EN ISO 527-4 [17] and UNE EN ISO 179-1 [18] (Fig. 2). The characterization of tensile mechanical properties (Tensile strength, σ_T, MPa and Young's Modulus, E_T, GPa) were carried out in an universal testing machine (Microtest EM2 300/FR), while the characterization of impact mechanical properties (Charpy impact strength, a_{CU}, kJ m^{-2}) were carried out in a Charpy pendulum for composites (Zwick I 5113.100).

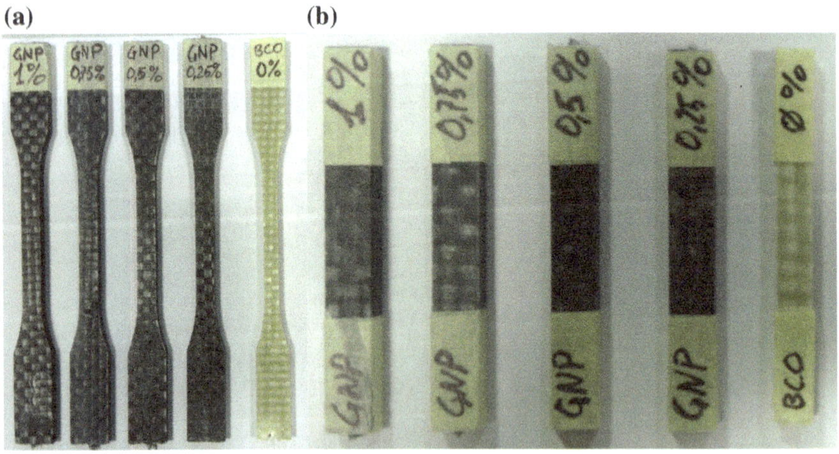

Fig. 2 Series of undoped (brown) and doped with graphene specimens (black) for the tensile tests (**a**) and for the Charpy impact tests (**b**)

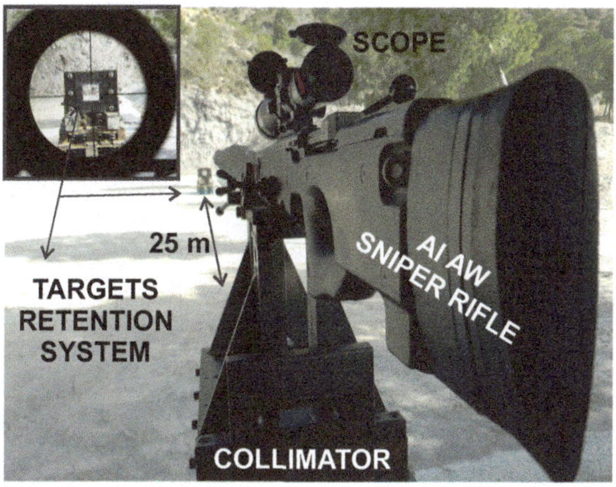

Fig. 3 Ballistic tests system consisting of the rifle, the collimator and the targets retention system. The shooting distance was 25 m. Top-left inserted picture shows the targets retention system seen through the scope and from the position of the shooter

2.4 *Ballistic Tests*

Shooting range tests were carried out in collaboration with the Spanish Navy and the Tercio de Levante of the Spanish Marine Corps. Characterization of the ballistic limits (V_0) of laminated plates were calculated from the experimental data of strike (V_s) and exit (V_e) velocities in accordance with the fitting algorithm (Eq. 1) established in the NATO STANAG 2920 standard [19]. For that purpose, a targets retention system was installed in the shooting range of this unit and an sniper rifle Accuracy International AW 07 11664 with 7.62 × 51 mm NATO FMJ ammunition (Pb/Sn/Sb core, Brass Jacket) was used as firing system (Fig. 3). The ammunitions workshop of the Algameca Naval Station provided for the experiments several series of cartridges with different propelling charges (100, 90, 80, 70, 60 and 50%), in order to obtain the ballistic behaviour of the materials in a wide range of impact velocities.

$$V_e^2 = V_s^2 - V_0^2 \tag{1}$$

3 Results and Discussion

3.1 *Characterization of FLG*

HR-TEM microscopy characterization of the few-layer pristine graphene used for this work is shown in Fig. 4. The images obtained allowed to observe the low number of layers in the FLG samples, to determine their average area, and to verify the distance between crystalline planes. The high value of average sheets area (2.47 ± 0.4 μm^2) and its small standard deviation demonstrated the monodispersion and the

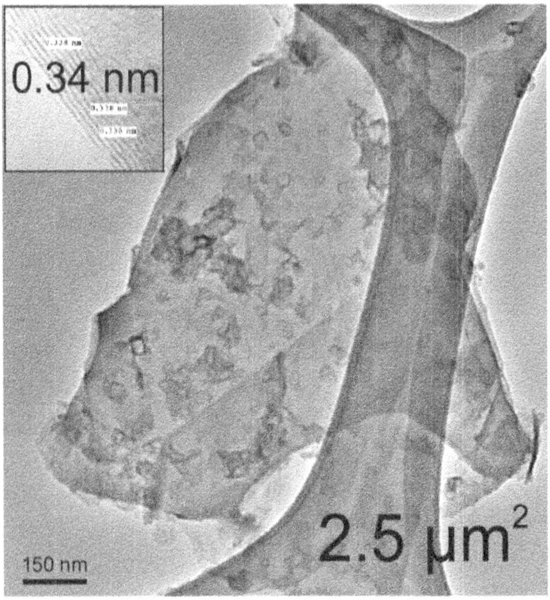

Fig. 4 HR-TEM image of FLG with its average area. The distance between crystalline planes is shown inserted (top-left)

high aspect ratio (thickness/area) of the obtained graphene (5.5×10^{-4}), which is an indicative parameter of its two-dimensional morphology. The distance between crystalline planes (0.338 nm) is practically the same as the theoretical one (0.34 nm) described in the bibliography [20].

Raman spectrum of the synthesized FLG and its three fundamental peaks (D, G and 2D) allowed to calculate its fundamental characterization parameters: I_D/I_G, I_{2D}/I_G, from which the distance between defects ($L_D = 17.44$ nm), the density of defects ($n_D = 1.06 \times 10^{-11}$ cm^{-2}) and the number of layers of the sample ($n \leq 7$) were obtained (Fig. 5), in accordance with the bibliography [21].

XRF elemental analysis composition of the produced FLG is shown in Table 1. Results obtained show that the material synthesized by our team for these experiments is, in effect, pristine few-layer graphene (3–7 layers) of a high quality.

3.2 *Mechanical Properties*

Experimental data obtained in the tensile tests allowed to obtain the stress-strain (σ/ε) curves, from which the ultimate breaking stress (σ_T) and the Young's Modulus (E_T) were calculated, as established in the corresponding standard [17] (Fig. 6).

Results obtained show that the addition of graphene produces, in most cases, an improvement in the tensile stress and Young's modulus with respect to the undoped laminate. These improvements present a maximum in both mechanical properties

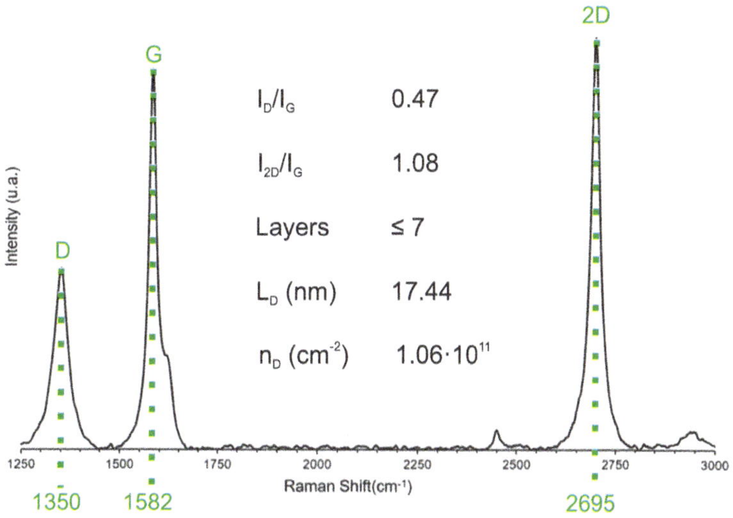

Fig. 5 Raman spectrum of FLG with the position of the peaks and its most relevant parameters

Table 1 XRF elemental analysis composition of the produced FLG

Element	Carbon	Chlorine	Silicon	Aluminium
(%)	99.29	0.52	0.14	0.05

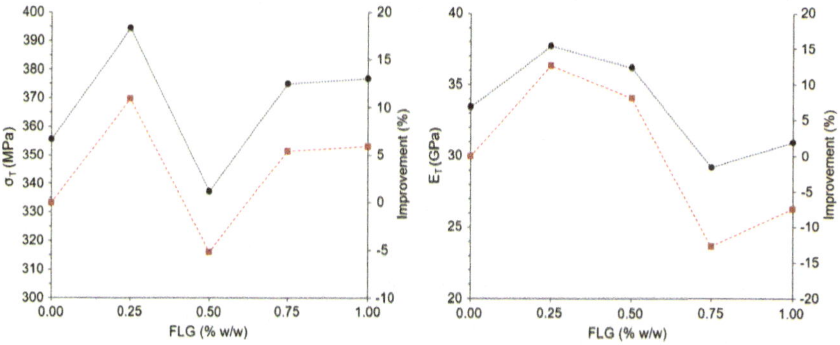

Fig. 6 (Left) Ultimate breaking stress (σ_T) and (right) Young's modulus (E_T) depending on FLG doping percentage with improvements regarding to the undoped samples (values in black and improvements in red)

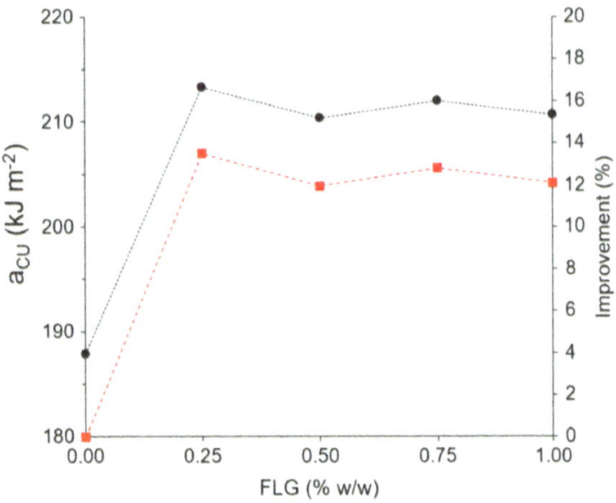

Fig. 7 Charpy impact strength (a_{CU}) depending on FLG doping percentage with improvements regarding to the undoped samples (values in black and improvements in red)

for the doping percentage of 0.25%, with relative values of 10.9% (394.4 MPa) and 12.7% (37.7 GPa), respectively.

Experimental data from the impact test allowed to obtain the values of absorbed energy in the fracture, from which the Charpy impact strength (a_{CU})were calculated, in accordance with it is established in the corresponding standard [18] (Fig. 7).

The addition of graphene produced an improvement in all doped samples with regard to the undoped laminate. As in the case of tensile properties, this improvement presents a maximum for the doping percentage of 0.25%, reaching a value of 13.5% (213.3 kJ m^{-2}).

3.3 *Ballistic Properties*

Experimental strike (V_s) and exit (V_e) velocities of the projectiles obtained in the real fire characterization tests allowed to determine the ballistic limit (V_0) of each sample, in accordance with the corresponding standard [19]. Figure 8 show the experimental data, the obtained values for the V_0 parameter and the improvements with regard to the undoped plate.

Results show that the V_0 parameter considerably improves with the increment of the FLG doping ratio, reaching a maximum value of 266.4 m s^{-1} for the 1% FLG doped laminate. This increase means a relative improvement of 72.2% with respect to the undoped laminate. Therefore, these results unequivocally support the viability of the development of new graphene-based nanocomposites with improved mechanical properties for the security and defence sector, both structural and, especially, ballistic protection.

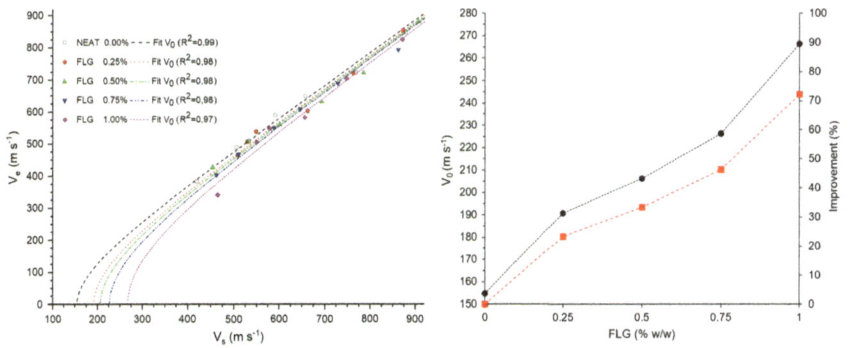

Fig. 8 (Left) Strike and exit velocities of tested plates with their fitting curves in accordance with Eq. 1. (Right) Ballistic limit V_0 obtained values with the relative improvement with regard to the undoped plate

4 Conclusions

In this work, the capability of graphene to improve the tensile, impact and anti-ballistic mechanical properties of conventional FGRP composites have been investigated, through the comparison of experimental data obtained testing undoped samples with those from several graphene-based nanocomposites doped in different percentages (0.25, 0.50, 0.75 and 1.00%) in weight.

The obtained results have confirmed that, in most cases, graphene acts as a nano-reinforcing agent of the base material in an effective way. The analysis of such data suggests that this effectiveness depends on the characteristics of applied force and the property studied, reaching maximum improvements of 10.9% in the tensile stress, 12.7% in Young's modulus, 13.5% in Charpy impact strength (0.25% FLG) and an outstanding 72.2% in the ballistic limit (1% FLG) against 7.62 × 51 mm NATO FMJ ammunition.

Tests carried out demonstrate the viability of developing new graphene-based nanocomposites with improved mechanical properties, which will allow the production of a much more efficient ballistic protection systems in a near future.

Acknowledgements The authors gratefully acknowledge the cofunding by the Spanish Ministry of Defence and the Technical University of Cartagena, under COINCIDENTE grant 1003215004200. Likewise, the authors want to express their gratitude to the Spanish Navy and to the Tercio de Levante of the Spanish Marine Corps, as well as the support provided by SAIT and the Materials Science and Engineering Department, both belonging to the UPCT.

References

1. Novoselov KS, Geim AK, Morozov SV, Jiang D, Zhang Y, Dubonos SV, Grigorieva IV, Firsov AA (2004) Science 306(5696):666
2. Varghese SS, Lonkar S, Singh K, Swaminathan S, Abdala A (2015) Sens Actuators B: Chem 218:160
3. Mohanty A, Srivastava VK, Sastry PU (2014) J Appl Polym Sci 131:1
4. Lee C, Wei X, Kysar JW, Hone J (2008) Science 321(5887):385
5. Kwolek S, Yang H (1993) Manmade fibers: their origin and development. Elsevier Applied Science Publishers Ltd., Barking, United Kingdom, pp 315–336
6. Kwolek SL (1971) Poly(p-benzamide) composition, process and product. US Patent 3,600,350
7. Twaron®. Aramid fiber. http://www.teijinaramid.com/aramids/twaron/
8. Dyneema®. UHMWPE fiber. http://www.dsm.com/products/dyneema/en_GB/product-technologies/fiber.html
9. Spectra®. UHMWPE fiber. http://www.honeywell-advancedfibersandcomposites.com/products/fibers/
10. Zylon®. PBO fiber. http://www.toyobo-global.com/seihin/kc/pbo/
11. Bracamonte L, Loutfy R, Yilmazcoban I, Rajan S (2016) In: Lightweight ballistic composites, 2nd edn. Elsevier, pp 349–367
12. Lee YS, Wetzel ED, Wagner NJ (2003) J Mater Sci 38(13):2825
13. Yamamoto N, Hart AJ, Garcia EJ, Wicks SS, Duong HM, Slocum AH, Wardle BL (2009) Carbon 47(3):551
14. Chou TW (2013) Advanced composite armor: in situ sensing with carbon nanotube networks for improved damage tolerance. Tech. rep., DTIC Document
15. Lee JH, Loya PE, Lou J, Thomas EL (2014) Science 346
16. Gago Velasco I, Ibarra Berrocal IJ, León Albert G, Miguel Hernández B (2016) ABS dopado con grafeno. Patent ES2570391B2
17. E. ISO (1997) Determination of tensile properties. Test conditions for isotropic and orthotropic fibre-reinforced plastic composites
18. U.E. ISO (2011) Determinación de las propiedades al impacto Charpy. Parte 1
19. N. STANAG 2920 (1999) Ballistic test method for personal armour materials and combat clothing
20. Hembacher S, Giessibl FJ, Mannhart J, Quate CF (2003) Proc Natl Acad Sci 100(22):12539
21. Cançado LG, Jorio A, Ferreira EM, Stavale F, Achete C, Capaz R, Moutinho M, Lombardo A, Kulmala T, Ferrari A (2011) Nano Lett 11(8):3190

Comparison of the Ballistic Stability of Composites Based on Carbide of Boron, Corund, and Glass-Ceramics (Armored Sitall)

Merkulov Yurii

Abstract This work shows the need to create a new light energy-intensive armored material to protect against bullets and fragments of modern ammunition. Comparative characteristics of modern ceramics, which are used for ballistic protection of people and equipment, are given. The ballistic stability (the limit of the rear strength) of composites based on boron carbide B_4C, Al_2O_3 corundum and developed high-strength glass ceramics based on lithium disilicate $Li_2Si_2O_5$ (armored glass) was compared. Ballistic tests confirmed the possibility of using armored glass in protective structures with a surface density of 35–38 kg/m^2 in the 5-th class of the RF standard P 50744-95/2014 and the 3rd class of the US NIJ standard. One of the options for reducing the surface density and the cost of composite armor when using high-strength glass ceramics (armored glass) is shown.

Keywords Armored sitall · Glass ceramics · Armor ceramics · Ballistic resistance · Damaging element · Rear durability limit

1 Introduction

The scientific and technological revolution taking place in the developed countries in the military sphere has led to the intensive development of armaments and military equipment. Along with the development of new, highly accurate and ever more "smart" types of combat equipment, there is also an active improvement of artillery and small arms: their rate of fire increases, the initial velocity of ammunition increases, their damaging effect increases, and their new forms are created. Artillery and small arms and in the development of missiles remain the most important means of achieving success in combat operations, ensuring the growth of combat losses of the opposing sides.

M. Yurii (✉)
Sosnovy Bor, Russian Federation
e-mail: melas33@yandex.ru

R. Fangueiro and S. Rana (eds.), *Advanced Materials for Defense*, 4, Springer Proceedings in Materials, https://doi.org/10.1007/978-3-030-34123-7_6

One of the ways to reduce combat losses is to provide servicemen with individual armored protection and to increase the stability of military equipment by engaging either the entire facility or its vital subsystems. The use of traditional steel armor for this purpose increases the mass of wearable equipment and greatly affects the fatigue and, as a consequence, the fighting efficiency of servicemen. In combat technology, the use of steel armor as the main defense significantly increases its mass and reduces many of its important tactical and technical characteristics: speed, load capacity, maneuverability, range, etc.

For example, protection against an armor-piercing bullet type B-32 caliber of 7.62 mm, released from a rifle SVD at a distance of 100 m, today can provide armor steel in a thickness of more than 10 mm with a surface density of about 90 kg/m^2. Therefore, it is quite natural (and not for the first time already) that scientists are faced with the task of creating a new light, energy-intensive and inexpensive armor to protect servicemen and vital subsystems of military equipment samples from their destruction by bullets and fragments of modern ammunition. To date, without the use of ceramic materials, this task cannot be solved. A comparison of the basic properties of armored ceramics with a rolled homogeneous armor RHA is given in Table 1 [1].

Developed and used today in the Armed Forces of Russia as armored ceramics have either low cost at high density (corundum ceramics B-6 is sold at a price of 17–35 USD for 1 kg at a density of 3.82–3.95 g/cm^3), or high cost at low density (ceramics based on boron carbide costs more than 250 USD per 1 kg at a density of 2.50–2.55 g/cm^3). Ceramics of silicon carbide with a density of 3.10–3.25 g/cm^3 can be bought at a price of 150–180 dollars USA [2]. There is a great need to create ceramics with a density of less than 3.0 g/cm^3 and cost 10–20 dollars USA for 1 kg for wide use in composite armor. One of the solutions to the problem of creating light energy-intensive armor material is the developed structural material based on armored glass ceramics (armored sitall).

Table 1 Comparison of basic properties of armored ceramics with a rolled homogeneous armor

	RHA	Al_2O_3 (aluminum oxide)	SiC (silicon carbide)	TiB_2 (titanium diboride)	B_4C (boron carbide)
Bulk density (g/cм^3)	7.85	3.81–3.92	3.09–3.23	4.45–4.52	2.50–2.52
Young's modulus (GPa)	210	350–390	380–430	520–550	420–460
Vickers hardness (kgf/mm^2)	300–550	1500–1900	1800–2800	2100–2600	2800–3400
Elongation before destruction (%)	14–18	<1	<1	<1	<1

According to McMillan's definition [3], glass-ceramic is a polycrystalline solid obtained by controlled crystallization. Crystallization is carried out by carefully controlled heat treatment, whereby crystallization centers are formed in the amorphous glass and crystal growth begins.

Armored sitall, which was developed by the author [4, 5], was obtained by controlled catalyzed crystallization of glass of a certain chemical composition, produced from ordinary silicate raw materials. From a large number of different options for designing high-strength glass-ceramics in terms of price-quality ratio (properties), the Li2O-ZnO-SiO2 system compositions were chosen. They have small differences in the composition, melting and heat treatment of glass, properties and cost (Table 2). LAS is an abbreviation for "light armored sitall", LS-33 is the serial number of one of the most successful variants of a light caliber.

2 Materials and Methods

The armored sitall used in the study has the following main characteristics [6]:

- melts temperature of the original glass: 1250–1300 °C
- crystallization temperature not more than 750 °C
- the density of glass-ceramic (sitall): 2.46–2.50 g/cm^3
- temperature coefficient of linear expansion: (110–125) × 10^{-7} K^{-1}
- relative hardness: 7.5–8 units according to Mohs (corundum level)
- strength at static bending of 350–450 MPa (35–45 kg/mm^2)
- shrinkage after crystallization no more than 3%
- withstand repeated thermal shock when cooled in water (from 500°C)
- The cost of raw materials in the Russian Federation is not more than $4. USA for 1 kg (retail).

Samples are composite plates of the "sitall-glass-plastic" type and consist of glass-reinforced support plates 100 × 100 × 10 mm and 200 × 200 × 10 with high-strength, 10 × 50 × 50 mm ceramic elements glued to them in the center. The following main components were used for sample preparation:

- Sitalls of grades LS33 and LAS, made by the author of work;
- Fiberglass of STET-1 brand, epoxy resin of brand ED-20.

In addition to these samples, to compare the protective characteristics in the test program, samples containing boron carbide (B_4C) ceramics and B-6 corundum ceramics (Al_2O_3) were included in the front layer. As a small weapon in the tests, the Zenit-3 rifle was used, and as ammunition—cartridges 7.62 × 54 with armor-piercing incendiary bullets B-32. As a comparison criterion, the rear durability limit was chosen. At present, there is no information on the use of glass ceramics in the

Table 2 Difference in chemical composition between LS33 and LAS

Mass %	SiO_2 (Silicon oxide)	Li_2O (Lithium oxide)	K_2O (Potassium oxide)	MnO/MnO_2 (Manganese oxide)	ZnO/ZnO_2 (Zinc oxide)	$NaPO_3$ (Sodium phosphate)	CaF_2 (Calcium fluoride)	LiF (Lithium fluoride)	Na_2O (Sodium oxide)
LS33	55.0–75.0	10.0–20.0	1.0–3.0	0.1–5.0	0.1–5.0	1.0–5.0	0.5–3.0	1.0–10.0	–
LAS	55.0–65.0	10.0–20.0	1.0–3.0	0.1–5.0	0,1–5,0	1.0–5.0	0.5–3.0	1.0–10.0	0.1–10.0

Russian Federation as armor material. An insignificant number of published reference data on the protective characteristics of modern armor materials in tabular form or in the form of empirical dependencies does not allow using them in design engineering calculations. There are also no computational methods that allow to foreseen the protective characteristics of armored barriers with glass ceramic elements with sufficient accuracy. For these reasons, the resistance of prototypes to the effect of a given damaging element was determined experimentally.

The calculation of the parameters of the protective structure, i.e., the determination of the necessary thicknesses of all its components, can be performed in the first approximation, provided that the protective characteristics of the materials used are available. Protective characteristics are the dependencies connecting the reservation parameter with the interaction criterion for a particular striking element (bullets, fragment). As a reservation parameter, either the thickness of the barrier (h) or its surface density (ps) is used. In the role of the interaction criterion, the velocity of the rear durability limit (VRDL) or the throughput penetration rate (VTPR) is most often used.

The limiting speed of through penetration of the VRDL for an armored barrier of a given thickness is such a value of the velocity of the penetrating body, in which it penetrates the barrier and has a zero or close velocity behind its rear surface. At the VRDL speed, the maximum residual deformation is observed on the rear surface of the barrier, but still without disrupting the continuity of the rear surface. The experience of experimental shooting shows that it is more convenient to use the VRDL criterion, since it has a smaller spread of values for the same material than the VTPR.

As an argument for these dependencies, the surface density of the obstacle, ps, is the mass of one square meter of the armor tested. Depending on the values, the corresponding VRDL (or VTPR) index is determined experimentally. The obtained indices are applied to the VRDL-ps, coordinate grid and connected by a smooth line, for which the values of the coefficients of its approximation are then determined. For characteristics of anti-bullet resistance, linear approximation of the type is more often used:

$$\mathrm{VRDL} = \mathrm{K} * \mathrm{ps} + \mathrm{C},$$

where

K is a coefficient of proportionality;
Ps is the surface density of the obstacle;
C is the speed offset.

Experimentally, the speed of the dorsal strength limit is determined by the nature of the interaction with the barrier of the damaging element. The nature of the interaction is assessed visually according to five main criteria:

- non-penetration, i.e. The case when no residual deformation is observed on the rear surface of the sample;
- penetration—the case when there is a violation of the continuity of the rear surface by the core of the bullet or its fragment;
- VRDL (speed of the rear durability limit)—when a very slight disruption of the continuity of the rear surface in the form of hair-thin cracks or single discontinuities is observed on the rear surface at maximum residual deformation;
- Non-penetration close to VRDL—in the case of large residual strains, but without disrupting the continuity of the rear surface of the sample (for glass and organoplasty—whitening of the back layer at the impact site of more than 5 gauges of the damaging element);
- penetration close to VRDL—in the case of rupture of individual strands of the back layer of the sample when the sample-witness is not penetrated at a distance of 20 mm.

In accordance with the adopted methodology, a program of ballistic tests was developed, during which the determination of the bullet velocity was carried out with each shot (by a non-contact method, using inductive sensors). Visually determined the speed of the backstop of the rear strength of VRDL (for the three main criteria—penetration, non-penetration, VRDL). Then the surface density of the samples was calculated for the possibility of subsequent comparison with the results of dynamic tests of such structures. Based on the results obtained, a decision was taken on the corresponding change in the speed of the bullet, which was produced by changing the mass of the powder in the cartridge.

3 Results and Conclusions

The results of comparative ballistic tests are shown in Fig. 1. It can be seen that the developed high-strength glass ceramics is inferior in quality to ceramics from boron carbide (B_4C), but superior to corundum ceramics B-6 (Al_2O_3). Therefore, it can be assumed that such glass ceramics can be used as a front layer of armored composites for protection against bullets and debris for use in protective structures in the 5th class of the RF standard P 50744-95/2014 and the 3rd class of the US NIJ standard.

In Fig. 2 [1] it was shown that the developed glass ceramics LS33 and LAS, which were used for testing, are close in price to corundum ceramics B-6, but much cheaper than boron carbide ceramics.

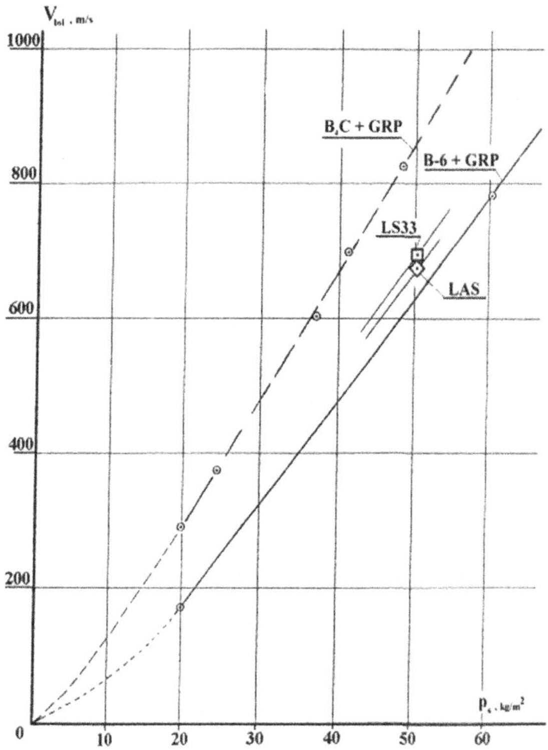

Fig. 1 Ballistic test results

Currently, armored panels with a surface density of not more than 40 kg/m^2 and cost not exceeding \$400 per m^2 are being developed for use in the 6a class of the RF standard (4th class of the NIJ standard). The next step is the creation of transparent glass ceramics to protect the optics of space stations [7].

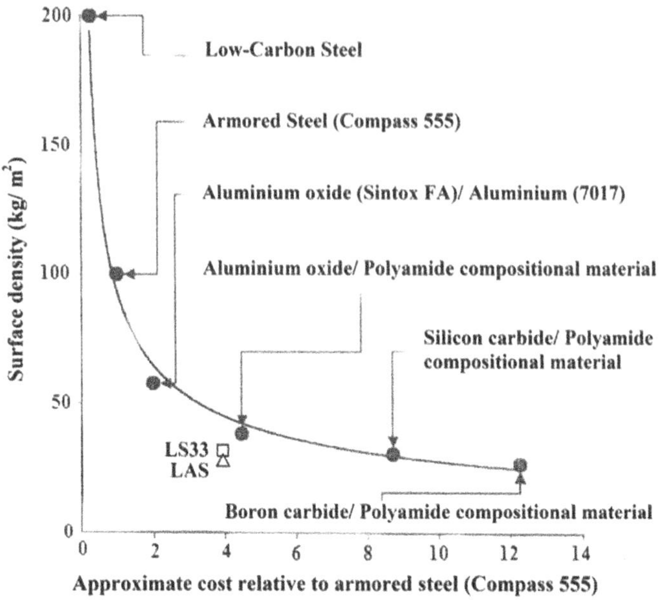

Fig. 2 Surface density of various composites required to protect against 7.52 mm armor piercing bullets, compared to their relative cost

Acknowledgements The author sincerely thanks Viktor Ilyukhin and Valery Vasyunkin for the assistance provided and the very opportunity to do this work.

References

1. Hazell PO (2009) Advances in ceramic armour military technology, vol XXXIII(4):118–126
2. Bespalov I, Grigoryan V, Karpov A (2012) Ceramic armor [electronic document]—news VPK 26(443). http// vpk-news.ru/ articles/ 9011
3. McMillan PW (1964) Glass-ceramics-AP. London, New York, p 261
4. Patent for invention—Merkulov YY (2001) Glass ceramics, a method for obtaining it and a protective construction based on it—Pat. Russian Federation No. 2176624, 2001, BI No. 34
5. Patent for invention—Merkulov YY (2007) Composite material and method of making composite material (variants)—Pat. RF No. 2294517, 2007, BI No. 6
6. Merkulov YY (2007) Armor sitall—new materials for ballistic protection. In: Materials of the 3rd international conference "technical means of countering terrorist and criminal explosions", St. Petersburg, pp 86–93
7. Merkulov YY, Sol'k SV, Lebedev OA (2017) Protection of the optics of the orbital station from high-speed damaging elements. Acta Astronaut 135:21–25

UHPFRC Resistance to Projectile Impact in Dependence on Fibre Content

Jindřich Fornůsek, Michal Mára, Radoslav Sovják and Jan Zatloukal

Abstract Ultra-high-performance fibre-reinforced concrete (UHPFRC) was subjected to projectile impact using in-service bullets. UHPFRC is an advanced cementitious composite with enhanced mechanical properties where high tensile strength and ductility benefit mainly from the disperse fibre reinforcement. Steel micro-fibres used in this study were straight and smooth with a tensile strength of 2800 MPa. The fibre content in the UHPFRC was set as the main test variable in the framework of this study. It was found that an increase in the fibre volume fraction resulted in an increase in both the tensile strength and the energy dissipation capacity. Cylinder compressive strength of the UHPFRC was determined to be 150 MPa. The UHPFRC slabs used as targets against full metal jacket soft lead core (FMJ-SLC) and full metal jacket mild steel core (FMJ-MSC) projectiles were 50 mm and 45 mm thin, respectively. Both projectiles were fired from the semi-automatic rifle. The weight of the projectiles was 8.04 g and muzzle velocity was determined to be 710 m/s. Damage parameters such as depth of penetration, crater diameter, residual velocity, residual penetration potential and debris fragment loss were determined. Based on the experimental observations it was found that the optimal fibre content in the UHPFRC mixture with respect to damage parameters induced by the high-speed projectile impact is about 2% by volume.

Keywords Fibre reinforced concrete · Projectile impact · Fibre content

1 Introduction

Current word events clearly showed that the level of protection of existing structures needs to be significantly improved. Hence a primary loading type that needs to be addressed is the localized impact of projectiles. Here, the projectile impact can be considered as the high-speed impact from an object having relatively a low weight. This might be a projectile from the direct armed attack and also projectile generated

J. Fornůsek (✉) · M. Mára · R. Sovják · J. Zatloukal
Experimental Centre, Faculty of Civil Engineering, Czech Technical University in Prague, Thákurova 7, 166 29 Prague, Czech Republic
e-mail: jindrich.fornusek@fsv.cvut.cz

R. Fangueiro and S. Rana (eds.), *Advanced Materials for Defense*, 4, Springer Proceedings in Materials, https://doi.org/10.1007/978-3-030-34123-7_7

from an explosion or from a failure of high-speed rotating machines. Several damage mechanisms are activated as once during projectile perforation, such as compaction on the proximal face, multiaxial stress state during penetration and tension at the distal face of the impacted element. The complexity of the process is further increased due to the fact, that the strain rate is different for every aforementioned damage mechanism. The goal of this study is to describe the damage to Ultra-High-Performance Fibre-Reinforced Concretes (UHPFRCs) with various fibre volume fractions that are impacted by various projectiles.

2 UHPFRC

Ultra-high-performance fibre-reinforced concrete (UHPFRC) can be classified as a composite material with high potential for protection and defence structures [5]. The UHPFRC is characterised by strain hardening behaviour which is reflected by enhanced fracture energy [6]. Strain hardening behaviour of the UHPFRC benefits from the disperse fibre reinforcement. Fibres used in this study were 13 mm long with a diameter equal to 0.15 mm. Fibres were straight and smooth with a tensile capacity of 2800 MPa, as specified by the manufacturer. These fibres provided a good trade-off between workability of the fresh mixture and mechanical properties of the hardened concrete. The UHPFRC used in this study was developed with respect to the components available commonly in the Czech Republic. The material design process was fully described elsewhere [3, 4]. Briefly, the UHPFRC was mixed in the conventional vertical mixers and the slabs were cured in the water tanks. The UHPFRC was blended using high cement and silica fume content, a low water-binder ratio and absence of coarse aggregate, i.e. aggregate larger than 1 mm. Straight steel micro-fibres were used in the UHPFRC mixture in order to improve the ductility of the material. In this study, the fibre content, as the most expensive part of the UHPFRC [8], was selected as the main test variable.

3 Mechanical Properties

Unconfined compressive strength and Young modulus of elasticity were measured on cylinders with a diameter of 100 mm and height of 200 mm. The strength of the best available capping material (100 MPa) was significantly lower than the expected measured strengths, consequently, the tops of the cylinders were cut off and grinded. Compressive strength was measured on cylinders by monotonic increments of pressure with an average rate of 36 MPa/min up to 70% of the expected compressive strength. At this point, loading was switched to deformation controlled loading with a rate of 0.48 mm/min for about 2 min in order to measure peak and post-peak behaviour. In the softening branch, the loading rate was increased to 1.2 mm/min.

Modulus of elasticity was measured using two strain-gauges with a 100 mm base, attached to the sides of the cylindrical specimen. A hydraulic loading machine with a maximal capacity of 2500 kN was used and the loading procedure was stress controlled. In the first step the specimens were loaded to 1/3 of the expected maximal compressive strength—in this case, 50 MPa—for 60 s. Afterwards, the specimens were unloaded to 5 MPa. This procedure was repeated three times. The Young modulus of elasticity was calculated from the third unloading branch. In the second step, the specimens were loaded until failure and compressive strength was determined.

Direct tensile tests were carried out on the dog-bone shaped specimens without a notch. The direct tensile tests were performed on MTS loading machine. The specimens were mounted into specially developed grips. The extension in the elastic region was measured with two strain-gauges glued on both narrow sides. After the localization of a crack, the extension was measured with two LVDTs mounted with a special frame on the dog-bone specimen. The loading rate was 0.2 mm/min.

Flexural strength in the three-point bending configuration was measured on the 100 × 100 × 400 mm large beams with a clear span of 300 mm. The loading rate was set to 0.2 mm/min. Fracture energy was determined on the 100 × 100 × 550 mm large beams with a notch and clear span of 500 mm. During the test deflection in the mid-span of the specimen was recorded. The fracture energy was determined based on the load-deflection curve as specified by RILEM recommendations [1].

4 Experimental Program

The UHPFRC used in this study was subjected to a series of tests that simulated the impact of the high-speed projectile with an impact velocity of 710 m/s. The penetration resistance was examined with a deformable missile with a full-metal jacket and soft-lead core (FMJ-SLC) and a non-deformable missile with a full-metal jacket and mild-steel core (FMJ-MSC) projectiles. Both projectiles were fired from a semi-automatic rifle against the UHPFRC slab with dimensions of 300 × 400 mm. Both projectiles are approved for civilian use.

The striking kinetic energy was approximately 2000 J. Projectiles used in this study were chosen in order to simulate high-speed low-weight fragments generated by an explosion, failure of the rotating machine or possibly by a direct armed attack caused by the small firearms.

The shooting distance was 20 m and the shooting chronograph to measure the actual muzzle velocity was located approximately 1 m from the gun muzzle (Fig. 1). The weight of both ogive-nose shaped steel jacket projectiles was 8.04 g and the diameter of both projectiles was 7.92 mm. The length of the FMJ-SLC and FMJ-MSC projectile was 23.3 mm and 26.7 mm, respectively [7].

Each test specimen was placed in a special mount developed for the purpose of this study in order to prevent movement of the specimen during the impact. The mount was provided with eight screws located in the corners of the specimen approximately 50 mm from both edges in order to simulate point supports. As the presence of

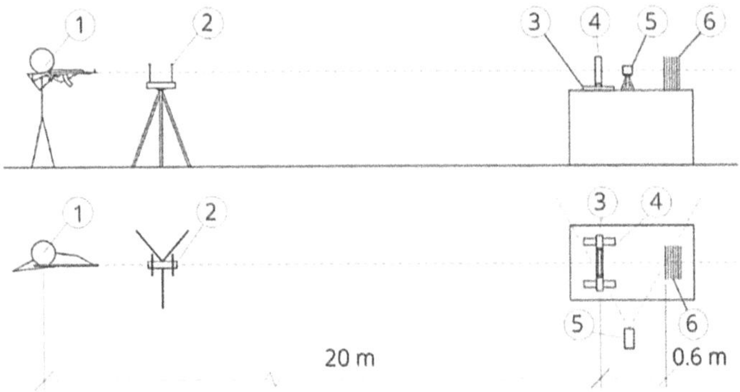

Fig. 1 Experimental setup of the projectile impact test; 1-shooter, 2-shooting chronograph, 3-mount for the slab, 4-UHPFRC slab, 5-high-speed camera, 6-set of 1 mm thin witness steel plates

reinforcing bars has a negligible effect on perforation resistance of the slabs, no reinforcing bars were utilized to construct the specimens.

A high-speed camera was placed behind the rear face of the UHPFRC slabs in order to capture the residual velocity of the projectile. A set of 1 mm thin witness steel plates was placed 0.6 m behind the rear face of the slab in order to evaluate the residual penetration potential of the bullet. The perforation of 1.5 plates is comparable to a life-threatening event for a human body [2].

5 Results and Discussion

Experimental readings of mechanical properties of the developed mixtures with respect to the fibre content are listed in Table 1 including compressive strength, flexural strength (modulus of rupture), tensile strength, Young modulus of elasticity and effective fracture energy. The values presented in Table 1 are averages from at least five samples. It can be seen that the average highest compressive strength of 151.7 MPa was achieved for 2% volumetric content of fibres. In correspondence with

Table 1 Mechanical properties of the UHPFRC mixtures

	UHPFRC			
Fibre content (% vol.)	0	1	2	3
Workability—spread (mm)	200	220	225	190
Compressive strength (MPa)	132.4	148.5	151.7	148.1
Flexural strength (MPa)	9.9	27.0	40.1	47.5
Direct tensile strength (MPa)	6.6	7.8	9.9	10.9
Modulus of elasticity (GPa)	41.1	45.1	56.9	51.5
Effective fracture energy (kJ/m^2)	0.35	12.0	17.9	25.3

Table 2 FMJ-SLC projectile impact

	UHPFRC			
Fibre content (% vol.)	0	1	2	3
Depth of penetration (mm)	Perf.	20.5	20.0	19.0
Crater diameter (mm)	148	85	74	73

that, the highest average secant modulus of elasticity of 56.9 GPa was measured for the mixture containing 2% of fibres by volume. The highest fracture energy, flexural and direct tensile strength was achieved for samples containing 3% of fibres by volume. The mixture containing 3% of fibres per volume displayed worse workability, which probably resulted in a decrease of some mechanical properties.

The semiautomatic rifle CZ 858 used in this study was calibre 7.62 × 39, which is a civilian version of a former Czechoslovak military assault rifle SA-58. The rifle was chosen because it represents the most commonly used firearm in the world, thus it sufficiently guarantees a level o response to a real attack. The experiment was arranged in order to achieve normal impact.

The results of the deformable FMJ-SLC projectile impact tests including the depth of penetration and the crater diameter are presented in Table 2. The deformable FMJ-MSC missile consisted of a lead body and steel jacket. Perforation through the slab occurred only in the case of the plain concrete specimen (i.e. with no disperse fibre reinforcement). In case of the fibre reinforced concrete, the slabs were not perforated and the projectile bounced back. The average depth of penetration was determined to be 20.5, 20.0 and 19.0 mm for fibre reinforced concrete slabs with 1%, 2% and 3% of fibre content, respectively. It is therefore evident that the increase of the fibre content above 1% has a negligible effect on the depth of penetration.

The damage caused to the front and rear side of the slab made of the plain concrete specimen was characterised by several macro cracks that propagated through the specimen to the rear side and by radial cracks that propagated from the centre to the edges, splitting the specimen into several pieces. The crater diameter was measured to be 148 mm for the plain concrete specimen. With the implementation of the fibres into the mixture, the crater diameter was decreased significantly and the overall damage was reduced. The average crater diameter was measured to be 85, 74 and 73 mm for slabs with 1%, 2% and 3% of fibre content, respectively. In addition, no macro cracks were observed on the specimen after the projectile impact test.

The results of the FMJ-MSC projectile impact tests including residual velocity, residual penetration potential and debris fragment mass are presented in Table 3. All slabs were perforated and the residual velocity is presented as a percentage of the impact velocity. The residual velocity was determined from the high-speed camera recording with 4000 frames per second. It was also confirmed that normal perforation occurred and neither pitch nor yaw angle was recorded during experimentation. The impact velocity was 710 m/s for all samples.

Experimental results showed that significant improvement was achieved when the fibre content was increased from 1 to 2% by volume. Within this interval, the

Table 3 FMJ-MSC projectile impact

	UHPFRC			
Fibre content (% vol.)	0	1	2	2.5
Residual velocity (%)	37	32	9	7
Residual penetration potential (−)	3	2.5	0	0
Debris fragment mass (g)	560	99	98	126

residual velocity dropped by 23%. When comparing the UHPFRC with 2% of fibres by volume to their plain concrete counterparts, the residual velocity dropped by 28%. Further increase of fibre content beyond 2% displayed a little improvement, however, the most significant contribution was achieved within 2% of fibre content by volume.

The positive influence of the fibre content on the residual velocity of the projectile was additionally confirmed by the number of perforated witness plates. A number of perforated witness plates pointed on the residual penetration potential of the projectile. The perforation of 1.5 witness plates is comparable to a life-threatening event for a human body [2]. Less than 1.5 perforated witness plates on average were gained only within the UHPFRC mixtures with 2.0 and 2.5% of fibre content by volume.

Implementation of relatively small fibre content enhanced the behaviour in terms of debris fragment mass significantly. Secondary fragments, i.e. fragments generated by the scabbing of the slab, could be as dangerous as primary fragments attacking the slab. Significant reduction of the debris fragment mass is evident within the interval between plain mixture and mixture containing 1.0% of fibres. The debris fragment mass was reduced from 560 g (plain mixture) to 99 g (UHPFRC mixture with 1.0% of fibre content). Further increase of fibre content beyond 1% did not bring any significant improvement in terms of reduction of the debris fragment mass.

6 Conclusions and Further Outlook

A series of ballistic experiments were carried out where UHPFRC was subjected to a number of blows from in-service projectiles under the normal impact. The addition of the high strength steel fibres to the mixture enhanced the impact behaviour of the UHPFRC slabs in terms of all damage parameters presented in this study, compared to their plain concrete counterparts. It was demonstrated that the optimal amount of the fibre content in the UHPFRC developed in the framework of this study is about 2% by volume. The steel fibres significantly contributed to the reduction of the damage of the UHPFRC specimens. Increase in fibre content resulted in improvement of shear angle; i.e. angle of stress distribution. This effect resulted in a larger volume of the material resisting the point impact, lower local stressing and higher penetration resistance. This kind of material provides the structure with essential resistance while maintaining its standard thickness and appearance because it has much greater ability to dissipate mechanical energy and to absorb shock loads. Implementation of the

UHPFRC may result in highly resistant concrete elements such as cladding panels and walls in modern protective structures while maintaining its standard thicknesses and appearance.

Acknowledgements The authors gratefully acknowledge the financial support from the Ministry of Industry of the Czech Republic [project number FV10547]. The authors also acknowledge assistance from the technical staff at the Experimental Centre, Faculty of Civil Engineering, Czech Technical University in Prague; and students who participated in the project.

References

1. Bažant ZP, Kazemi MT (1991) Size dependence of concrete fracture energy determined by RILEM work-of-fracture method. Int J Fract 51:121–138. https://doi.org/10.1007/BF00033974
2. Bludau C, Keuser M, Kustermann A (2006) Perforation resistance of high-strength concrete panels. ACI Struct J 103
3. Maca P, Zatloukal J, Konvalinka P (2012) Development of ultra high performance fiber reinforced concrete mixture. In: IEEE symposium on business, engineering and industrial applications (ISBEIA), IEEE, pp 861–866
4. Máca P, Sovják R, Konvalinka P (2013) Mix design of UHPFRC and its response to projectile impact. Int J Impact Eng 63:158–163. https://doi.org/10.1016/j.ijimpeng.2013.08.003
5. Millon O, Riedel W, Mayrhofer C, Thoma K (2010) Fiber-reinforced ultra-high performance concrete—a material with potential for protective structures. In: Li QM, Hao H, Li ZX, Yankelevsky D (eds) Proceedings of the first international conference of protective structures, Manchester, p 13
6. Sovják R, Rašínová J, Máca P (2014) Effective fracture energy of ultra-high-performance fibre-reinforced concrete under increased strain rates. Acta Polytech 54:358–362. https://doi.org/10.14311/AP.2014.54.0358
7. Sovják R, Vavřiník T, Zatloukal J et al (2015) Resistance of slim UHPFRC targets to projectile impact using in-service bullets. Int J Impact Eng 76:166–177. https://doi.org/10.1016/j.ijimpeng.2014.10.002
8. Wille K, Boisvert-Cotulio C (2015) Material efficiency in the design of ultra-high performance concrete. Constr Build Mater 86:33–43. https://doi.org/10.1016/j.conbuildmat.2015.03.087

Mechanical Properties of Ultra High Performance Fibre Reinforced Concrete and Its Response to Impact Loading

Anna L. Mina, Konstantinos G. Trezos and Michael F. Petrou

Abstract This manuscript presents the findings of an experimental study aiming to develop an Ultra High Performance Fibre Reinforced Concrete (UHPFRC) best suitable for impact loading. The mechanical properties of this material and the failure mechanisms under impact loading were investigated. The UHPFRC developed in this research program is a cement-based material with increased percentage of microsilica and a combination of two types of steel fibres: 6 mm long with 0.16 mm diameter and 13 mm long with 0.16 mm diameter. It includes local calcarenite sand with grains 125–500 μm in size. Four composition mixtures were produced, three UHPFRC and one Ultra High Performance Concrete (UHPC), without steel fibres. The influence of the amount and type of steel fibres was examined. Slabs with dimensions: 200 × 200 mm and thickness of 15, 30, 50 and 70 mm were produced and were subjected to projectile impact (real shotgun used). The penetration depth and material volume loss were measured.

Keywords High performance concrete · Steel fibres · Impact loading · Penetration depth · Volume loss

1 Introduction

In recent years, the increase of terrorist attacks creates the need to design more secured structures, leading to new types of concrete. UHPC is a special type of concrete that shows higher compressive strength than normal concrete. The disadvantage of high strength concrete is that it becomes more brittle. To overcome this disadvantage, steel fibres with high tensile strength are added and Ultra-High Performance Fibre Reinforced Concrete (UHPFRC) is produced. Steel fibres control cracking and offer increased toughness. UHPFRC with advanced mechanical properties can be used for reinforced concrete structures which are subjected to impact loadings [1, 2].

A. L. Mina (✉) · K. G. Trezos
School of Civil Engineering, National Technical University of Athens, Athens, Greece
e-mail: mina.anna@ucy.ac.cy

A. L. Mina · M. F. Petrou
Department of Civil and Environmental Engineering, University of Cyprus, Nicosia, Cyprus

R. Fangueiro and S. Rana (eds.), *Advanced Materials for Defense*, 4, Springer Proceedings in Materials, https://doi.org/10.1007/978-3-030-34123-7_8

This study aims to develop an UHPFRC with mechanical properties specifically designed for projectile impact [3]. Three mixtures were composed and their mechanical properties were studied. Specifically, the compressive strength, the direct tensile strength, the flexural tensile strength, the modulus of Elasticity and the Poisson's ratio were measured. Finally, slabs with dimensions: 200 × 200 mm and thickness of 15, 30, 50 and 70 mm were produced and tested under projectile impact. Slab penetration depth and material volume loss were measured and the failure mechanisms were investigated.

2 Experimental Program

2.1 *Mixtures and Test Specimens*

The mixture design is based on Benson and Karihaloo [4] which was modified by Nicolaides et al. [3], in order to use local materials from Cyprus. For this study, four compositions were produced, from which one was UHPC, without steel fibers and the other three were UHPFRC, with the addition of steel fibers. Two types of steel fibres were used: one type was 6 mm long and 0.16 mm diameter and the other type had the same diameter 0.16 mm but was 13 mm long. The composition of the mixtures is the same, with a differentiation regarding the quantity and type of the steel fibers. As it shown in Table 1, one mixture of UHPFRC included the smallest amount of 2% per volume 13 mm long fibers. The other two mixtures of UHPFRC contained both types of steel fibers with total amount of 6% per volume.

The labels of the mixtures are represented as follows: a ($a_1 - a_2$)/b/c where a, a_1, a_2, b and c are: percentage (%) of the fibres by volume, percentage (%) by volume of the fibres 6 mm long, percentage (%) by volume of the fibres 13 mm long, the percentage (%) of the water to binder ratio and the percentage (%) of the super plasticizer to binder ratio, respectively.

The materials used to prepare the mixtures are presented below:

Table 1 Compositions of mixtures

Materials (kg/m^3)	0/16/6	2(0-2)16/6	6(5-1)/16/6	6(3-3)/16/6
Cement	880	880	880	880
Microsilica	220	220	220	220
Steel fibres 6 mm	0	0	400	240
Steel fibres 13 mm	0	160	80	240
Sand 125–250 μm	475	475	475	475
Sand 250–500 μm	358	358	358	358
Water	176	176	176	176
Superplasticizer	67	67	67	67

Table 2 Type and dimensions of specimens for each test

Type of specimens and dimensions	Cubes (a_1 = a_2 = a_3 = 100 mm)	Cylinder (d = 100 mm, h = 200 mm)	Prismatic beams (a_1 = 100 mm, a_2 = 50 mm, l = 500 mm)	Prismatic beams (a_1 = 100 mm, a_2 = 100 mm, l = 500 mm)	Slabs (a_1 = a_2 = 200 mm, t = (15/30/50/70)mm
Test	Compression	Compression	Direct tension	Indirect tension	Projectile impact

- Portland Cement CEM I strength category 52,5R
- Microsilica with a silica content (SiO_2) 97%
- Polycarboxylate polymer based on superplasticiser
- Calcarenite sand from local quarry, grading 125–250 and 250–500 μm
- Steel fibres 0.16 mm diameter and 6 mm or 13 mm length.

The mixing of the constituent materials was made in a vertical axis mixer and all specimens were prepared in a single batch. The specimens intended for measuring the mechanical characteristics of each mixture were prepared according to the CYS EN12390-1:2009 Standard. The specimen dimensions and intended testing procedures are presented in Table 2.

2.2 Experimental Tests for Mechanical Properties

Cubical specimens were utilized to measure the compressive strength and cylindrical specimens to measure the ultimate compressive strength with corresponding strain, Modulus of Elasticity and Poisson's ratio. The compression tests were conducted using an automatic 5000 kN hydraulic compression machine. Cubes were tested according to CYS EN 12390-3:2009, at a rate of 0.5 MPa/s. For uniaxial compression test [5], a displacement rate of 0.015 mm/min was used (Fig. 1a).

Direct tension tests on prismatic beams were conducted using special stainless steel grippers with hinged heads (Fig. 1b), according to a technique adopted from Kolias and Williams [6]. Direct tension tests were performed with a 300 kN materials testing machine and a displacement rate of 0.04 mm/min. Ultimate tensile strength and associated strain were recorded.

Indirect tension test, known as three point bending test on prismatic beams, was conducted using a hydraulic pump machine according to EN 12390-5:2009 (Fig. 1c). Displacement and flexural load were measured.

Strain measurements were obtained using strain gauges and displacements using LVDTs. Strain gauges were fixed along the specimens and LVDTs were attached on rings that embraced the specimens

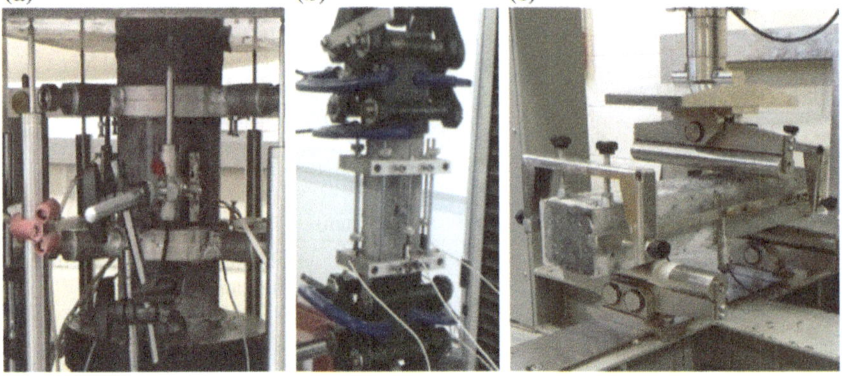

Fig. 1 Tests set—up of **a** compression, **b** direct tension and **c** indirect tension

2.3 Experimental Test for Projectile Impact Load

Firearm G3, with 7.62 mm bullets (Fig. 2a), was used to examine the performance of slab specimens under impact loading. Wooden bases (Fig. 2b) were constructed to support the slab specimens in the field and were placed at a distance of 100 m from the firing position. For the specific firearm, the highest velocity of the bullet is 800 m/s. At a distance of 100 m, no velocity decrease is expected. Eight slabs were constructed from each mixture with four different thicknesses 15, 30, 50 and 70 mm (2 slabs for every thickness). The target surface of the slabs was 200 × 200 mm.

During impact loading there are three possibilities (Fig. 3):

- The bullet passes completely through the slab (failure type I)
- The bullet penetrates into the slab but is unable to go completely through it and causes a loss of material in a cone shape in the front face, cracks may appear in the rear face (failure type II)
- The slabs are destroyed by breaking into pieces (failure type III).

Fig. 2 **a** Bullet of the firearm G3 **b** Slab setup in the field

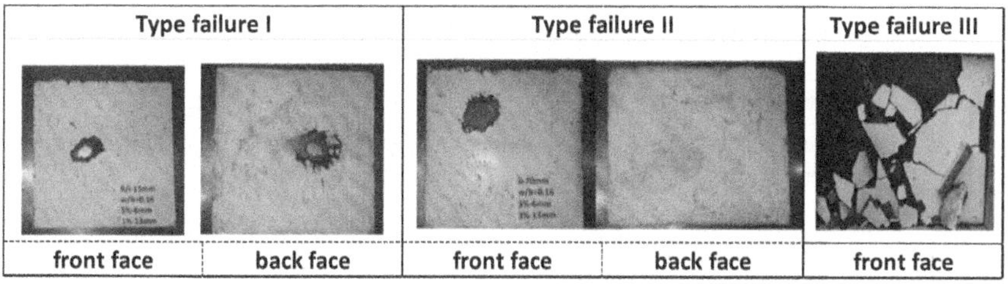

Fig. 3 Types of failure

The slabs were categorized based on the three types of failure (Fig. 3a–c) after the impact load. In addition, the penetration depth and material volume loss were measured.

3 Results and Disscusion

In Tables 3 and 4, the mechanical properties of the mixtures are presented.

The addition of steel fibres, 2% per volume, improved the behaviour of the mixture. The compressive strength raised by 25.6% for cubes and 3.0% for cylinders. Also, the associated strain increased by 7.3%. The compressive strength was further increased by 13.4% when the volume of steel fibers raised from 2 to 6% (ratio of short steel fibers to long 5:1) and by 6.2% when equal amount of short and long fibres were included in the mixture included (3% of each type of fibres). The associated strain exhibited a significant increase of 30.4% with the increase of the fibres from 2 to 6% (short to long fibres ratio 5:1). The Modulus of Elasticity E, increased with the addition of steel fibers in the mixture, and the mixture 6(3-3)/16/6 achieved the highest value of 38.6 GPa. The Poisson's ratio did not show any significant change. The mixture 6(3-3)/16/6 exhibited the best mechanical properties based on the compression tests.

Table 3 Compression strength with corresponding strain, modulus of elasticity and Poisson's ratio

Compression test					
Mixture	Compression strength from cubes (MPa)	Compression strength from cylinders (MPa)	Corresponding strain × 10^{-3} of maximum compressive strength	Modulus of elasticity E (GPa)	Poisson's ratio ν
0/16/6	124.0	131.3	3.98	33.1	0.25
2(0-2)/16/6	155.8	135.3	4.27	35.2	0.22
6(5-1)/16/6	176.7	158.4	5.57	34.2	0.26
6(3-3)/16/6	187.6	167.1	5.36	38.6	0.25

Table 4 Direct tension strength with corresponding strain and indirect tension strength with corresponding mid span deflection

Mixture	Direct tension test		Indirect tension test	
	Direct tensile strength from beams (MPa)	Corresponding strain × 10^{-3} of maximum direct tensile strength	Indirect tensile strength (MPa)	Deflection (mm)
0/16/6	4.4	0.12	13.2	0.08
2(0-2)/16/6	9.2	0.83	26.6	0.72
6(5-1)/16/6	9.5	0.21	31.7	0.53
6(3-3)/16/6	6.5	0.20	34.2	0.86

The increase of steel fibres by 2% per volume in the mixture increased the direct tensile strength by 109% and the indirect tensile strength by 101%. A small increase of the ultimate stress of 3.3% was noticed in direct tension and a significant increase of 19.2% in indirect tension, with the increase of steel fibres from 2 to 6% (short to long steel fibres ratio 5:1) by volume. Long fibers helped the development of high strain in direct tension. In indirect tension the substitution of short length fibres with long ones helped to increase by 62.2% of the displacement value. In the next Tables 5, 6 and 7 the results from the projectile impact testing are shown. All slabs from mixture 0/16/6 had type III failure and broke into pieces. It was not possible to measure penetration depth and material volume loss.

As is shown in Tables 5 and 6, the mixture with 2% steel fibres exhibited complete penetration in all slabs of 15 and 30 mm thickness and in one of 50 mm thickness. When the percentage of steel fibres increased to 6% (5:1 ratio for short to long fibers), complete penetration appeared in slabs with 15 and 30 mm thickness. With the substitution of short length steel fibres of 2% per volume with long length fibres, mixture 6(3-3)/16/6, complete penetration was only observed in slabs of 15 mm thickness.

Table 5 Failure type for mixtures 0/16/6 and 2(0-2)/16/6

Mixture 0/16/6			Mixture 2(0-2)/16/6		
Dimensions of slab (mm)	Thickness of slab (mm)	Failure type	Dimensions of slab (mm)	Thickness of slab (mm)	Failure type
200 × 200	15	III	200 × 200	15	I
200 × 200	15	III	200 × 200	15	I
200 × 200	30	III	200 × 200	30	I
200 × 200	30	III	200 × 200	30	I
200 × 200	50	III	200 × 200	50	I
200 × 200	50	III	200 × 200	50	II
200 × 200	70	III	200 × 200	70	II
200 × 200	70	III	200 × 200	70	II

Table 6 Failure type for mixtures 6(5-1)/16/6 and 6(3-3)/16/6

Mixture 6(5-1)/16/6			Mixture 6(3-3)16/6		
Dimensions of slab (mm)	Thickness of slab (mm)	Failure type	Dimensions of slab (mm)	Thickness of slab (mm)	Failure type
200 × 200	15	I	200 × 200	15	I
200 × 200	15	I	200 × 200	15	I
200 × 200	30	I	200 × 200	30	II
200 × 200	30	I	200 × 200	30	II
200 × 200	50	II	200 × 200	50	II
200 × 200	50	II	200 × 200	50	II
200 × 200	70	II	200 × 200	70	II
			200 × 200	70	II

Table 7 Average penetration depth and material volume loss for each mixture, for 70 mm thickness slabs

Mixture	Thickness of slabs (mm)	Average penetration depth Pd (cm)	Average material volume loss (cm^3)
2(0-2)/16/6	70	1.63	18.4
6(5-1)/16/6		1.69	14.9
6(3-3)/16/6		0.82	10.0

For 70 mm thickness slabs, there was no failure in any of the mixtures. Measurements of the average penetration depth and material volume loss are presented in Table 7. The increase of steel fibres from 2 to 6%, 5:1 ratio short to long steel fibres, resulted in a decrease of volume of material by 19.0% and had minor effect on penetration depth. The substitution of short length fibres of 2% per volume with long length fibres, mixture 6(3-3)/16/6, reduced the volume loss by 32.9%. There was also a mentionable decrease in the penetration depth by 51.5%.

4 Conclusions

Several conclusions can be drawn from the conducted research:

1. Additions of steel fibres improved the mechanical properties of UHPC and its resistance to impact loading.
2. UHPFRC slabs with thickness of 50 mm or higher, compressive strength of over 150 MPa and containing 6% per volume fibres exhibited no penetration.
3. The mixture with a composition 6(3-3)/16/6 exhibited the best mechanical characteristics and showed complete perforation under projectile impact only in 15 mm thickness slabs.

References

1. Zhang MH, Shim VPW, Lu G, Chew CW (2005) Resistance of high-strength concrete to projectile impact. Int J Impact Eng 31:825–841
2. Maca P, Sovjak R, Konvalinka P (2014) Mix design of UHPFRC and its response to projectile impact. Int J Impact Eng 63:158–163
3. Nicolaides D, Kanellopoulos A, Petrou M, Savva P, Mina A (2015) Development of a new ultra high performance fibre reinforced Cementitious composite (UHPFRCC) for impact and blast protection of structures. Constr Buildings Mater 95:667–674
4. Benson BSDP, Karihaloo L (2005) CARDIFRC—development and mechanical properties. Part I: development and workability. Mag Concr Res 57(6):347–352
5. Michael. FP, Konstantinos GT, Mina AL (2015) Design of UHPFRC mixtures to be used in structures subjected to impact loads. In: 27th Biennial national conference of the concrete Institute of Australia in conjunction with the 69th RILEM week, Melbourne, Australia, 30 Aug–02 Sept
6. Kolias S, Williams RIT (1978) Uniaxial tension tests on cement-stabilized granular materials. Geotech Test J GTJODJ 1(4):190–198

A Reduced-Order Modeling Based Design and Optimization for a Lightweight Multilayer Armor Plate Against Blast and Impact

Weiran Jiang, Alyssa Bennett, Nickolas Vlahopoulos and Geng Zhang

Abstract The modeling based design and optimization has been widely used in many engineering fields to avoid the high cost of extensive experiments. A reduced-order model, which mainly considers the critical degrees of freedom in the design, is even more efficient and effective in significantly saving computational efforts without losing the design accuracy and optimization reliability. The information generated from the reduced-order model in the early design stages then will be a useful guideline for further comprehensive analysis in a full-order model. A new reduced-order modeling method has been developed for evaluating the dynamic performance of a lightweight multilayer armor plate subjected to blast and impact loadings. Based on the reduced-order model, a fast optimization framework is introduced in this paper and utilized as a computational efficient tool for rapidly identifying the optimal design from a large number of candidate configurations. A simplified human model seated on the multilayer plate and its lumbar injury criteria is employed as a target screening metric. Four design cases with a combination of steel, SiC, Al-Alloy, Ti-Alloy, and copper are analyzed, compared, and validated by a commercial finite element analysis package. A good design balance between structural weight reduction and protection capacity enhancement has been achieved by using this optimization framework.

Keywords Reduced-order modeling · Fast optimization · Lightweight structure · Multilayer plate · Composite materials · Blast mitigation · Impact resistance

1 Introduction

Occupant centric vehicle structures that provide safety from blast and impact threats while at the same time reduce the weight of the vehicle structure, comprises one of the main thrusts in the US Army Science and Technology (S&T) activities. Such

---

W. Jiang (✉) · A. Bennett · N. Vlahopoulos
Department of Naval Architecture and Marine Engineering, University of Michigan, Ann Arbor, USA
e-mail: leaduwin@umich.edu

G. Zhang
Michigan Engineering Service LLC, Ann Arbor, USA

R. Fangueiro and S. Rana (eds.), *Advanced Materials for Defense*, 4, Springer Proceedings in Materials, https://doi.org/10.1007/978-3-030-34123-7_9

structures that provide a design balance between lighter weight and better survivability under increasingly severe blast threats, such as mines and improvised explosive devices (IED), are highly desirable.

One of the new designs of absorbing the destructive energy is to utilize the properties of softer structural materials in combination with a damping mechanism that can be used for absorbing the energy through deformation. These softer materials are able to reduce the shock loads that are transferred in the interior of the vehicle by absorbing more energy through higher deformation than that of characteristic of normal high strength materials. Through the tuning of material properties, the structural components made of the softer materials can be utilized to their full advantage as an energy absorber and isolator to enhance the survivability of the vehicle in blast events [1, 2].

Another new design is to utilize anisotropic material properties to guide and redirect the destructive energy away from the occupants along pre-designated energy paths [3–5]. However, few monolithic materials and structures appear likely to be able to meet the above expectations and handle the conflicting demands of greater survivability and lighter structural weight. Thus, innovative functional-graded materials and multilayer armor plate structures have been proposed and developed. The dynamic performance of multilayer armor plates is of great interest because they act as a mechanism to absorb and spread the energy from a blast load in the lateral direction instead of permitting it to enter occupant space [6, 7]. This concept is promising but it is also extremely costly not only in setting up experiments, but also in modeling and computing if we use the general finite element analysis in parametric study and design optimization for a large number of alternative design configurations.

A reduced-order modeling approach has been developed recently for rapidly evaluating the structural dynamic performance of multilayer plates [8]. This reduced-order model (ROM), which is built based on the reverberation matrix method and generalized ray theory from the perspective of wave propagation in solids, is able to rapidly generate information for identifying the preferred spreading directions and amounts of energy transfer [9–16]. In combination with a spinal injury probability based screening metric, the ROM can be modified and used as an efficient tool to evaluate the performance of thousands of candidate multilayer plates. To identify the multilayer plate that has the highest levels of blast and impact protection and lightest structural weight among a large number of alternative configurations, a design optimization process is developed by integrating the ROM with it.

Based on the ROM, a framework of design optimization is introduced in this paper and a set of numerical case studies are provided for identifying the optimal configuration with different number of sublayers. First, an effective screening metric is developed, which employees the human lumbar injury criteria as a design target [17–19]. This screening metric will help to rank candidate configurations according to their capacity of blast isolation and impact protection. Then a framework of design optimization, including its objective function, design constrains, and candidate material selection, is developed and applied by employing the NSGA-II optimization solver [20, 21]. Four case studies (a three-layer case, a four-layer case, a five-layer case, and a six-layer case) are presented and the optimal configurations are then identified with the smallest screening metric at a reduced structural weight. The selected configuration for each case is validated by using finite element analysis in Nastran. The information generated from the design optimization will help designers to select an optimal multilayer plate configuration as a proper armor in an early design stage, in order to reduce the occupant's injury probability and the structural weight simultaneously. Then the identified optimal configurations can be used in a more complex and computational expensive finite element analysis.

2 The Reduced-Order Multilayer Material Model

The reverberation-ray matrix method, which is a combination of the reverberation matrix method and the generalized-ray method, is employed to develop the ROM for analyzing the dynamic response of a multilayer material under an external excitation. The multilayer material can be treated as a discrete functional graded material model with different sublayers. The material properties of each sublayer are pre-defined in this ROM, including modulus of elasticity E_i, Poisson ratio ν_i, mass density ρ_i, and thickness of sublayer t_i. The ROM is a two-dimensional model. To apply the ROM, a general infinite multilayer plate with N layers and N + 1 interfaces $y = y^J$ (J = 0, 1, 2, …, N), is considered and shown in Fig. 1. A two-dimensional coordinate system (*x, y*) is constructed in which the wave normal of the p-wave (in-plane longitudinal wave) and s-wave (shear wave) lie. The double local coordinate system $\left(x^{IJ}, y^{IJ}\right)$ and $\left(x^{I(I-1)}, y^{I(I-1)}\right)$ at each interface and a regular local coordinate system at the top surface boundary $\left(x^{01}, y^{01}\right)$ and the bottom surface boundary $\left(x^{N(N-1}, y^{N(N-1)}\right)$ are defined and shown respectively in Fig. 1.

The material properties, including mass density ρ^J, Lame's elasticity modulus λ^J and μ^J, and the thickness h^J, are isotropic in each sublayer but can be varying between different layers. It's worth noting that the Lame's elasticity moduli are derived as $\lambda = \frac{E*\nu}{(1+\nu)(1-2\nu)}$ and $\mu = \frac{E}{2(1+\nu)}$ respectively. In this paper, all the loads are assumed from external excitations and no loading sources exist at interfaces or within sublayers. The excitation applied at the interface and/or at the top and bottom boundaries is defined as a loading source vector $\mathbf{f}^J$. Only the propagation of elastic waves is investigated in this ROM. The p-wave potential $\phi^{IJ}(x, y, t)$ with p-wave speed C_p^{IJ} and s-wave potential $\psi^{IJ}(x, y, t)$ with s-wave speed C_s^{IJ} are defined in each sublayer. Both wave potentials satisfy classical wave equation:

$$\nabla^2\phi = \frac{1}{c_p^2}\frac{\partial^2\phi}{\partial t^2} \tag{2.1}$$

$$\nabla^2\psi = \frac{1}{c_s^2}\frac{\partial^2\psi}{\partial t^2} \tag{2.2}$$

where, P-wave speed is $c_p = \sqrt{\frac{\lambda+2\mu}{\rho}}$; S-wave speed is $c_s = \sqrt{\frac{\mu}{\rho}}$;

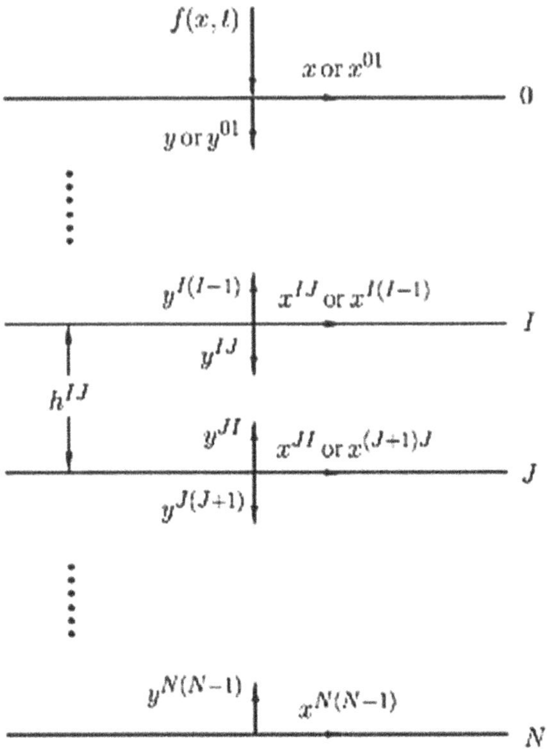

Fig. 1 A multilayer plate and its local coordinate systems at the boundaries and interfaces

By applying the double Fourier Transform on the fundamental wave equations in the space and time domains, the wave equations are represented in the wave number k and frequency ω domains:

$$\frac{d^2\tilde{\phi}}{dy^2} + \alpha^2\tilde{\phi} = 0 \tag{2.3}$$

$$\frac{d^2\tilde{\psi}}{dy^2} + \beta^2\tilde{\psi} = 0 \tag{2.4}$$

where the wave numbers are:

$$\alpha = \sqrt{\frac{\omega^2}{c_p^2} - k^2} : \text{(p-wave number in y direction);}$$

$$\beta = \sqrt{\frac{\omega^2}{c_s^2} - k^2} : \text{(s-wave number in y direction);}$$

The solutions for the wave equations can be expressed in terms of the arrival wave amplitude and departure wave amplitude as follows:

$$\tilde{\phi}(k, y, \omega) = \tilde{a}_p(k, \omega)e^{-i\alpha y} + \tilde{d}_p(k, \omega)e^{i\alpha y} \tag{2.5}$$

$$\tilde{\psi}(k, y, \omega) = \tilde{a}_s(k, \omega)e^{-i\beta y} + \tilde{d}_s(k, \omega)e^{i\beta y} \tag{2.6}$$

In Eqs. (2.5) and (2.6), $\tilde{a}_p, \tilde{a}_s$ are the unknown arrival wave amplitudes and $\tilde{d}_p, \tilde{d}_s$ are the unknown departure wave amplitudes that need to be solved. Once the wave amplitudes vectors are obtained, the corresponding displacement field and the stress components can be solved as well. The displacement fields $u = \left[u_x, u_y\right]$ and the stress components $\sigma = \left[\sigma_{xy}, \sigma_{yy}\right]$ can be calculated by applying fundamental elastodynamic solutions in each local coordinate system:

$$u_x = \frac{\partial \phi}{\partial x} - \frac{\partial \psi}{\partial y} \tag{2.7}$$

$$u_y = \frac{\partial \phi}{\partial y} + \frac{\partial \psi}{\partial x} \tag{2.8}$$

$$\sigma_{xy} = \mu\left[2\frac{\partial^2 \phi}{\partial x \partial y} + \frac{\partial^2 \psi}{\partial x^2} - \frac{\partial^2 \psi}{\partial y^2}\right] \tag{2.9}$$

$$\sigma_{yy} = \lambda \nabla^2 \phi + 2\mu\left[\frac{\partial^2 \phi}{\partial y^2} + \frac{\partial^2 \psi}{\partial x \partial y}\right] \tag{2.10}$$

The scattering phenomenon at each interface is shown in Fig. 2, where ***a*** represents the "arrival waves" and ***d*** represents the "departure waves". Also, the subscript ***p*** represents the "p-wave", which is also called the longitudinal wave, and the subscript ***s*** represents the "s-wave", which is also called the shear wave.

During the scattering, the incident waves will separate into reflected waves and refracted waves, and the wave types will interconvert during scattering. Therefore, a local scattering matrix is built to represent this complicated phenomenon by linking the incident waves (arrival) to the transmitted/reflected waves (departure). Basically,

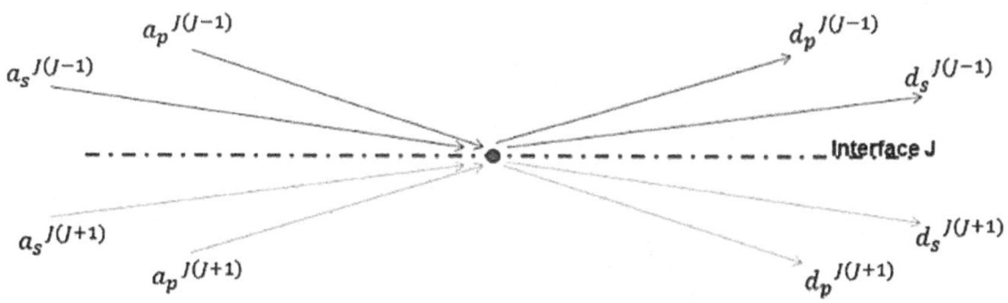

Fig. 2 The scattering phenomenon at interface J

if there are no wave sources inside the multilayer plate, and assume all of the adjacent layers share perfect interfaces without energy loss during the scattering, the displacement and stress yield the continuity conditions at each interface, as shown in Eqs. (2.11)–(2.14):

$$\tilde{u}_x^{J(J-1)}(k,\omega) - \tilde{u}_x^{J(J+1)}(k,\omega) = 0 \quad (2.11)$$

$$\tilde{u}_y^{J(J-1)}(k,\omega) + \tilde{u}_y^{J(J+1)}(k,\omega) = 0 \quad (2.12)$$

$$\tilde{\sigma}_{xy}^{J(J-1)}(k,\omega) + \tilde{\sigma}_{xy}^{J(J+1)}(k,\omega) = 0 \quad (2.13)$$

$$\tilde{\sigma}_{yy}^{J(J-1)}(k,\omega) - \tilde{\sigma}_{yy}^{J(J+1)}(k,\omega) = 0 \quad (2.14)$$

By substituting the elastodynamic solution Eqs. (2.7)–(2.10) into the above equations, a linear system is built to describe the scattering phenomenon, where $\boldsymbol{A}^J$ and $\boldsymbol{D}^J$ are both 4 × 4 matrices.

$$\boldsymbol{A}^J \cdot \tilde{\boldsymbol{a}}^J + \boldsymbol{D}^J \cdot \tilde{\boldsymbol{d}}^J = \tilde{\boldsymbol{g}}^J(k,\omega) = 0 \quad (2.15)$$

The linear system can be also written in the form of Eq. (2.16), by multiplying $\left(\boldsymbol{D}^J\right)^{-1}$ to Eq. (2.15):

$$\tilde{\boldsymbol{d}}^J = \boldsymbol{S}^J \cdot \tilde{\boldsymbol{a}}^J + \boldsymbol{s}^J \quad (2.16)$$

where,

$\boldsymbol{S}^J = -\left(\boldsymbol{D}^J\right)^{-1} \cdot \boldsymbol{A}^J$ is the local scattering matrix at interface J;
$\boldsymbol{s}^J = -\left(\boldsymbol{D}^J\right)^{-1} \cdot \tilde{\boldsymbol{g}}^J$ is the local source vector at interface J

In order to solve the two unknown wave amplitude vectors $\tilde{a}$ and $\tilde{d}$, another equation is necessary to be provided besides the Eq. (2.16). An additional equation is developed and supplemented by noting that a wave departing from one side of the layer becomes the wave arriving at another side of the same layer, with a phase lag, which is shown on next page in Eqs. (2.17)–(2.20).

$$\tilde{a}_p^{J(J-1)} = e^{i\alpha_J h_J} \cdot \tilde{d}_p^{(J-1)J} \quad (2.17)$$

$$\tilde{a}_s^{J(J-1)} = -e^{i\beta_J h_J} \cdot \tilde{d}_s^{(J-1)J} \quad (2.18)$$

$$\tilde{d}_p^{J(J-1)} = e^{-i\alpha_J h_J} \cdot \tilde{a}_p^{(J-1)J} \quad (2.19)$$

$$\tilde{d}_s^{J(J-1)} = -e^{-i\beta_J h_J} \cdot \tilde{a}_s^{(J-1)J} \quad (2.20)$$

Then similarly, all the "arrival waves" could be linked to the "departure waves" by a 4 N × 4 N global phase matrix $\boldsymbol{P}(h)$ as:

$$\tilde{\boldsymbol{a}} = \boldsymbol{P}(h) \cdot \tilde{\boldsymbol{d}}^{*} \quad (2.21)$$

The departure waves vector $\tilde{\boldsymbol{d}}^{*}$ and $\tilde{\boldsymbol{d}}$ contain the same elements but in different order. They can be expressed in equivalence through a global permutation matrix $\boldsymbol{U}$, which is only used to rearrange the order of the elements in the matrix. By combining the global scattering matrix, $\boldsymbol{S}$, global phase matrix, $\boldsymbol{P}$, and global permutation matrix, $\boldsymbol{U}$, the 4 N × 4 N reverberation matrix $\boldsymbol{R}$ is built as:

$$\boldsymbol{R}(k, \omega) = \boldsymbol{S} \cdot \boldsymbol{P} \cdot \boldsymbol{U} \quad (2.22)$$

The reverberation matrix can also help to link the unknown arrival wave amplitudes and the unknown departure amplitudes to the external excitation in the frequency and wave number domains.

Therefore, the arrival wave amplitudes are:

$$\tilde{\boldsymbol{a}} = \boldsymbol{P} \cdot \boldsymbol{U} \cdot [\boldsymbol{I} - \boldsymbol{R}]^{-1} \cdot \tilde{\boldsymbol{s}} \quad (2.23)$$

The departure wave amplitudes are:

$$\tilde{\boldsymbol{d}} = [\boldsymbol{I} - \boldsymbol{R}]^{-1} \cdot \tilde{\boldsymbol{s}} \quad (2.24)$$

The dynamic response of multilayer plate in terms of the displacement field can be calculated by substituting $\tilde{a}$ and $\tilde{d}$ into the elastodynamic equations and expressed in a matrix form as shown in Eq. (2.25):

$$\boldsymbol{W}(k, y, \omega) = \boldsymbol{A}_{\boldsymbol{u}} \cdot \tilde{\boldsymbol{a}} + \boldsymbol{D}_{\boldsymbol{u}} \cdot \tilde{\boldsymbol{d}} = (\boldsymbol{A}_{\boldsymbol{u}} \cdot \boldsymbol{P} \cdot \boldsymbol{U} + \boldsymbol{D}_{\boldsymbol{u}}) \cdot [\boldsymbol{I} - \boldsymbol{R}]^{-1} \cdot \tilde{\boldsymbol{s}} \quad (2.25)$$

The matrix A_u and D_u are the global receiver matrices, in which the elements are the coefficients in elastodynamic solutions that contain the user-defined receiver location information. By applying inverse Fourier Transform in wave number domain, the frequency response can be represented in space-frequency domain:

$$\boldsymbol{W}(x, y, \omega) = \frac{1}{2\pi} \int_{-\infty}^{\infty} \boldsymbol{W}(k, y, \omega) e^{ikx} dk \quad (2.26)$$

The frequency response at any locations of the multilayer plate, including the top and bottom surfaces and all the interfaces, can be calculated by using Eq. (2.26).

3 A Screening Metric for Evaluating Multilayer Armor Plates

Now, the vertical displacement at the center of the bottom surface of the multilayer plate can be obtained from the ROM. But the vertical displacement is not suitable to perform as an effective screening metric for determining which configuration has the best occupant protection capacity. For example, the injury probability of occupant in an underbody blast event highly depends on the deformation of the lumbar rather than the absolute displacement at the base of the structure. In this paper, the human lumbar deformation is employed as a high fidelity screening metric for evaluating the protection capacity of the multilayer armor plates based on their dynamic performance. A spinal system of human body consists of upper body, lumbar, and pelvis, where the upper body and the pelvis are connected by the lumbar. For the occupant seating on an armor plate subjected to an underbody explosive load, which is very similar to an impulse excitation, most of the blast energy is delivered to the lumbar in a short time period along with a significant lumbar compression, and only small amount of the blast energy is transmitted to the upper body of the occupant so that the deformation of the upper body can be negligible. Then it is suitable to simplify the entire upper body as a rigid-body lumped mass and the lumbar as a mechanism that links the upper body and the pelvis by a spring and a damper connected in parallel. As a result, a simple mass-spring-damper system is created to simulate the biomechanical response of the human spinal system, as illustrated in Fig. 3.

The biomechanical response is modeled as a linear second-order ordinary differential equation with lumbar compression as the variable. The time-dependent acceleration shown in Fig. 3 is the shock load delivered to the pelvis location through the vehicle structure by the blast event. Using this loading as the forcing function, the governing equation for the SDOF model is:

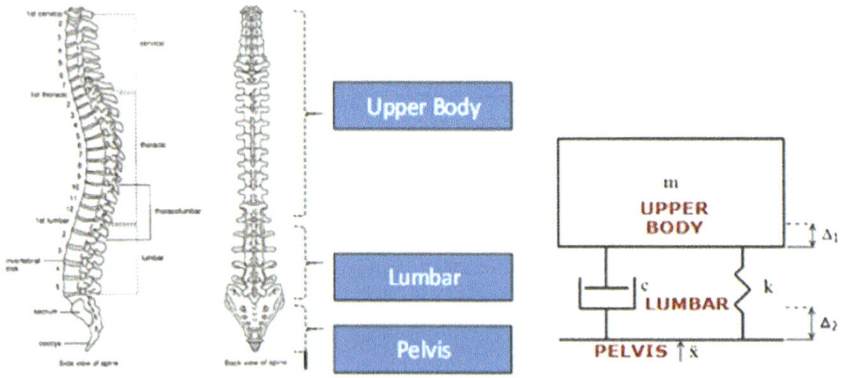

Fig. 3 A mass-spring-damper system for modeling the human spine system

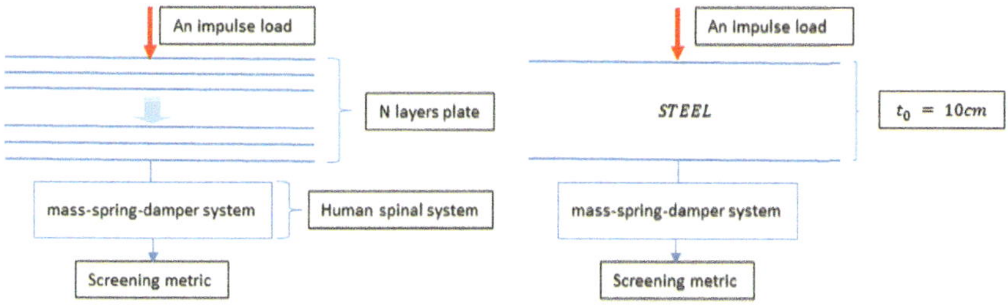

Fig. 4 The optimization and the baseline configuration

$$\frac{d^2\delta}{dt^2} + 2 \cdot \zeta \cdot \omega_n \frac{d\delta}{dt} + \omega_n^2 \cdot \delta = \frac{d^2x}{dt^2} \tag{3.1}$$

The parameter values of m, k, and c used here are derived by compressive strength of individual vertebrae and load-deflection curve from a representative population of Air Force pilots with a mean age of 28 years. The upper body lumped mass is approximated as 34.51 kg. A spring coefficient of 9.66×10^4 N/m and a damping coefficient of 818.1 N s/m are used to simulate the dynamic properties of the human lumbar. Therefore, the natural frequency ω_n and the damping ratio ζ of human lumbar are calculated as 52.9 rad/s and 0.224, respectively. Therefore, the lumbar compression δ, which is also the relative displacement between pelvis and upper body as $\delta = \Delta_2 - \Delta_1$ will be utilized to create the screening metric for effectively evaluating the occupant protection level of the multilayer armor plates. The screening metric is obtained by combining the ROM with the human spinal model, as shown in Fig. 4.

4 A Design and Optimization Framework

Having the screening metric for evaluating the energy insulation performance of the multilayer plate, a design optimization framework is developed to identify the optimal configuration. The configuration that needs to be optimized and a baseline configuration are shown in Fig. 4.

The optimization configuration has N sublayers with varying sublayer thickness. As discussed in preceding section, an impulse load is applied on the top surface of the multilayer plate to simulate a blast or impact event. A single degree-of-freedom mass-spring-damper system is attached at the center of bottom surface of the plate to capture the screening metric. The baseline model is a particular multilayer plate configuration with only one layer (N = 1), i.e. an isotropic plate, which is made of regular RHA high strength steel. The total thickness of the baseline plate model is set to be 10 cm. In the optimization, the total number of sublayers N, and the material selection for each sublayer s_j, are designated to be the two discrete design variables. The thickness of each sublayer t_i is designated to be a continuous design variable. The objective function of the optimization is to minimize the screening metric, which is calculated based on the largest relative displacement experienced by the mass-spring-damper system, as presented in Eq. (4.1):

$$\textbf{Objective Fucntion } f = \min(\text{Screening Metric}) \tag{4.1}$$

Four design constraints are also provided:

(1) To avoid the thickness of each sublayers being either excessive thick or extreme thin, a fixed upper bound of 4.8 cm and a lower bound of 1 cm are set on the thickness for each sublayer:

$$1\,\text{cm} < t_i < 4.8\,\text{cm} \tag{4.2}$$

(2) The total thickness of the multilayer plate is set to be unchanged to the total thickness of baseline model, with a 2% floating clearance:

$$9.8\,\text{cm} < \sum_i^N t_i < 10.2\,\text{cm} \tag{4.3}$$

(3) It's also important to keep the effective rigidity of the multilayer armor plate within an acceptable range to avoid dramatic deformation. Thus, the effective sectional stiffness is set to be greater than the one of the isotropic steel baseline configuration:

$$\sum_i^N E_i * t_i > 0.7 * (E_{steel} * t_0) \tag{4.4}$$

(4) Equally important to reducing the screening metric, reducing the overall structural weight is another crucial purpose of the design. Therefore, the total sectional density is set to be less than the one of the baseline in order to achieve the purpose of structural weight reduction:

$$\sum_i^N \rho_i * t_i < (\rho_{steel} * t_0) \tag{4.5}$$

The optimization is set up in DS Toolkit, which is a commercial package developed by Michigan Engineering Service (MES). The reduced-order model, which is implemented in MATLAB, is compiled and integrated with the DS Toolkit to build the optimization framework. The solver selected in the optimization is NSGA-II. The generation size is set to be 10 in order to ensure a numerical stable convergence for different configurations with varying number of sublayer in the following case study. The population size of each generation is set to be as large as 200, so that enough testing points are provided for identifying the feasible solutions. A group of

Table 1 Candidate materials for each sublayer and their corresponding material properties

# (s_j)	Materials selection	Modulus of elasticity (GPa)	Density (kg/m^3)	Poisson ratio
s_1	RHA steel	200	7823	0.30
s_2	6061 aluminum alloy	74	2704	0.34
s_3	SiC	445	3200	0.16
s_4	OFHC copper	129	8940	0.35
s_5	Ti-6AI-4V alloy	114	4430	0.34

five alternative materials are selected as candidates for each sublayer, including RHA high strength steel, 6061 Aluminum Alloy, Silicon-Carbon (SiC), OFHC copper, and Ti-6AI-4V-Alloy. The corresponding material properties are presented in Table 1.

5 The Optimal Design and Its Validation

In our case studies, four numerical testing configurations are prepared with different number of sublayers (N = 3, N = 4, N = 5, and N = 6). Both convergence rate and computational time increase when the number of sublayer increases. The optimal configurations of each testing case are shown in Fig. 5. The material selection is labeled for each sublayer. The sublayer thickness, total thickness, and the sectional density are also provided. It is observed that a variety of material combination is selected for each of the optimal configuration rather than having isotropic material for all of the sublayers.

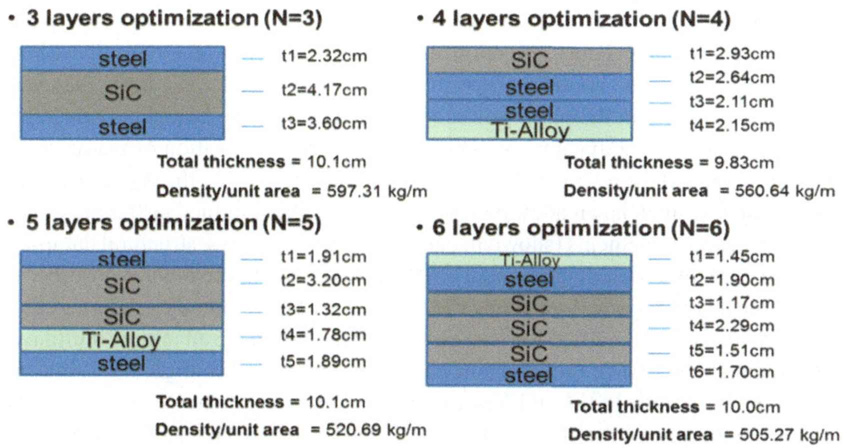

Fig. 5 The optimal configuration with different numbers of sublayers (N = 3, 4, 5, 6)

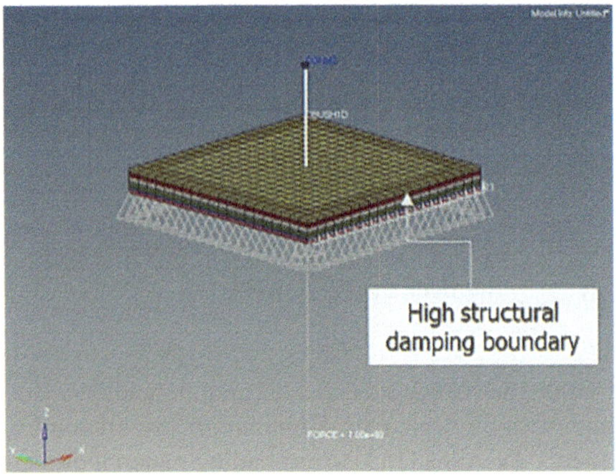

Fig. 6 A finite element model in Nastran for validating the optimal configurations

It is also observed that only three materials, including RHA high strength steel, SiC, and Ti-6AI-4V Alloy are selected from the five candidate material in Table 1 to compose the optimal configuration. To validate above optimization results, the classic finite element code in NASTRAN is used to analyze the optimal configuration that is identified when different numbers of sublayers are considered. The relative ranking of the optimal configurations that correspond to the different number of sublayers are evaluated and compared with the results obtained from the ROM based optimization. The Nastran model is shown in Fig. 6.

In this Nastran model, a 1 m × 1 m finite multilayer plate is created. An impulse load is applied at the center of bottom surface of the plate, and a single degree-of-freedom mass-spring-damper system is attached at the center of top surface of the plate. A simply-supported boundary condition is applied at four edges of the lower surface of the plate to prevent the mode of rigid body motion. Because in the optimization by utilizing the ROM, the multilayer plate model is an infinite plate, which indicates that, in lateral direction all of the waves will travel to infinity without any reflection from the boundary. However, the Nastran model is a finite plate and the waves travelling in the plate will be multi-reflected by the lateral boundary and result an accumulative dynamic response, which is a superposition of the response of direct wave and the response of reflected waves. To eliminate the difference from two unmatched models, an additional boundary is applied in the Nastran model by changing the outer elements (shown in darker colors) with a 40% structural damping. The high structural damping boundary will work as an effective wave absorber to prevent the incoming waves reflecting back when they hit the lateral boundaries of the plate. Therefore, the modified Nastran model is able to simulate an infinite multilayer plate to match the ROM. The screening metric of the optimal configurations that are calculated from both ROM and Nastran, are compared and shown in Fig. 7. The result from the baseline configuration is set to be 100%, and the results from optimal configurations are expressed accordingly in a percentage form.

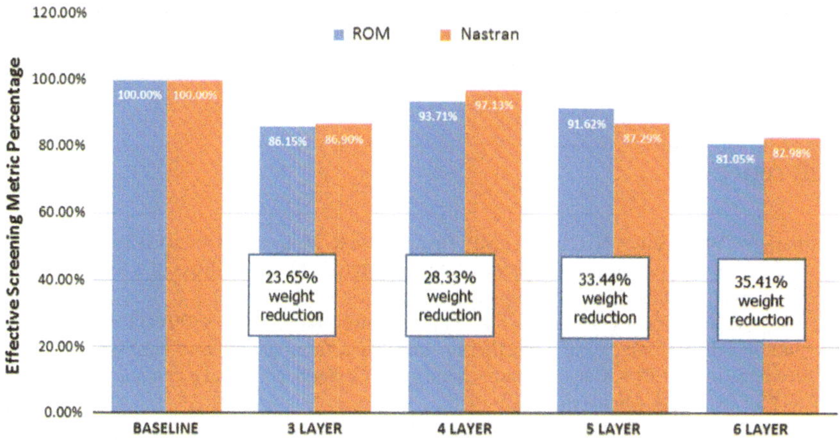

Fig. 7 The comparison between the results from ROM optimization and Nastran

It is observed that the screening metric calculated from the ROM optimization is similar to the results calculated from the Nastran model, which validates that this optimization framework is reliable for selecting lightweight multilayer armor plate with high protection ability. It is also found that by using the optimal six-layer plate configuration, both the screening metric and the weight of the structure have the greatest reduction. Based on current five candidate materials, the optimal six-layer plate configuration shown in Fig. 5 will be the best choice for a blast and impact isolation structure design to replace the isotropic steel plate.

6 Conclusion

This work presents an optimization framework for design lightweight multilayer plate configurations with high levels of blast and impact protection. An effective occupant centric screening metric is developed to better capture the levels occupant protection of the multilayer armor plate. The design optimization framework of a multilayer plate subjected to an impulse load is configured, for improving the screening matric while pursuing a structural weight reduction. The number of layers, the material selection, and the thickness of the layers comprise the design variables. The selected optimal configurations are validated by using finite element analysis in Nastran.

This efficient optimization process, which is based on the reduced-order multilayer plate model, provides a fast and reliable routine to identify the optimal multilayer plate among a large number of alternative configurations. It will help designers to select proper materials and determine the thickness of each sublayer in an early design stage of a multilayer plate armor design, which has a capacity to reduce the occupant injury probability and structural weight simultaneously. Moreover, by using the information generated from the ROM-based optimization, a higher-fidelity structural analysis can be performed in a more complex finite element model for relatively few configurations without performing a tedious parametric study.

References

1. Jiang W, Vlahopoulos N, Castanier MP, Thyagarajan R, Mohammad S (2015) Tuning material and component properties to reduce weight and increase blastworthiness of a notional V-hull structure. Case Stud Mech Syst Signal Process 2:19–28
2. Jiang W, Vlahopoulos N, Castanier M, Thyagarajan R, Mohammad S (2014) Reducing structural weight and increasing protection in simple structures subjected to blast loads. In: Proceedings of the 2014 NDIA ground vehicle systems engineering and technology symposium (GVSETS), Novi, Michigan, pp 12–14
3. Zhang Y, Hasan M, Starosvetsky Y, McFarland D, Vakakis A (2015) Nonlinear mixed solitary—shear waves and pulse equi-partition in a granular network. Physica D 291:45–61
4. Zhang Y, Moore K, McFarland D, Vakakis A (2015) Targeted energy transfers and passive acoustic wave redirection in a two-dimensional granular network under periodic excitation. J Appl Phys 118:234901
5. Zhang Y, Karr DG (2017) Determining ice pressure distribution on a stiffened panel using orthotropic plate inverse theory. J Struct Eng 04017003
6. Gupta YM, Ding JL (2002) Impact load spreading in layered materials and structures: concept and quantitative measure. Int J Impact Eng 27:277–291
7. Robbins JR, Ding JL, Gupta YM (2004) Load spreading and penetration resistance of layered structures—a numerical study. Int J Impact Eng 30:593–615
8. Jiang W, Bennett A, Vlahopoulos N, Castanier M et al (2016) A reduced-order model for evaluating the dynamic response of multilayer plates to impulsive loads. SAE Int J Passeng Cars Mech Syst 9(1):83–89
9. Pao YH, Gajewski R (1977) The generalized ray theory and transient response of layered elastic solids (Chap. 6). In: Mason WP (ed) Physical acoustics, vol X. Academic Press, New York
10. Pao YH, Su XY, Tian JY (2000) Reverberation matrix method for propagation of sound in a multilayered liquid. J Sound Vib 230(4):743–760
11. Chen WQ, Wang HM, Bao RH (2007) On calculating dispersion curves of waves in a functionally graded elastic plate. Compos Struct 81(2)
12. Su XH, Tian JY, Pao YH (2002) Application of the reverberation-ray matrix to the propagation of elastic waves in a layered solid. Int J Solids Struct 39(21–22)
13. Tian JY, Yang WX, Su XY (2006) Transient elastic waves in a transversely isotropic laminate impacted by axisymmetric load. J Sound Vib 289(1):94–108
14. Lee GS, Ma CC (2000) Transient elastic waves propagating in a multi-layered medium subjected to in-plane dynamic loadings I. Theory. Proc R Soc A 456(1998):1355–1374
15. Ma CC, Lee GS (2000) Transient elastic waves propagating in a multi-layered medium subjected to in-plane dynamic loadings II. Numerical calculation and experimental measurement. Proc R Soc A 456(1998):1375–1396
16. Zhang Y, Karr DG, Yu B (2016) Nonlinear contact dynamic response analysis of a matlock's ice-structure interaction model using fourier analysis. In: Proceedings of 23rd IAHR-ICE 2016, Ann Arbor, USA
17. Manion D (2013) Computation and validation of the dynamic response index (DRI). DTIC Report # ADA 591869, 6 Aug 2013
18. Stech EL, Payne PR (1969) Dynamic models of the human body. Aerospace Medical Research Laboratory (AMRL Technical Report 66–157), Nov 1969
19. Zimmermann RE, Merritt NA (1989) Aircraft crash survival design guide. U.S. Army Aviation Research and Technology Activity (USAAVSCOM TR 89-D-22A), Dec 1989

20. Deb K, Pratap A, Agarwal S, Meyarivan T (2002) A fast and elitist multiobjective genetic algorithm: NSGA-II. IEEE Trans Evolut Comput 6:182–197
21. Yusoff Y, Ngadiman MS, Zain AM (2011) Overview of NSGA-II for optimizing machining process parameters. Procedia Eng 15:3978–3983

Behind Helmet Blunt Trauma: Matching Measurement to Injury Using Finite Element Modelling

Ana Azevedo, Frederik Coghe and F. Teixeira-Dias

Abstract Nowadays, one of the risks related to the use of lightweight body armour systems is the risk on injury caused by a non-perforating impact. The dynamic deflection of the inner surface of the body armour can be sufficient to cause serious or even lethal injuries even if there is no full perforation of the body armour. The work presented here describes a methodology for an armour representative of an advanced ballistic helmet design and the human head. Using different experimental measurement methods, the traditional witness backing material method and more advanced methods using anthropomorphic test devices (ATD), reliable finite element models are created for the composite (poly-aramid and poly-ethylene) helmet shell. This is followed by numerical simulations where the interaction between the different armour solutions and the human body are studied in detail. Lastly, the results are linked to real-life injuries using the Abbreviated Injury Scale (AIS).

Keywords Blunt trauma · Injury · Ballistic helmet · Composites · Finite element modelling

1 Introduction

Personal armour systems frequently make use of fibre-based fabrics or composites to provide enhanced fragmentation and ballistic protection because of their high mass efficiency. Most legacy systems make use of poly-aramid fibre-based fabrics (e.g. DuPont Kevlar®, Teijin Twaron®, etc.). Relatively recently, poly-ethylene fibre-based solutions (UHMWPE) are being introduced, typically as uni-dimensional laminate (UD) solutions (e.g. DuPont Endumax®, DSM Dyneema®, Honeywell Spectra®, etc.). The use of UHMPWE in personal armour, such as helmets, tends to lead to lighter, and more mass-efficient armour systems. However, helmets with lower

A. Azevedo (✉) · F. Coghe
Department of Weapon Systems and Ballistics, Royal Military Academy, Avenue de la Renaissance 30, 1000 Brussels, Belgium
e-mail: ana.ferreira@mil.be

F. Teixeira-Dias
School of Engineering, The University of Edinburg, Edinburgh EH9 3JL, UK

R. Fangueiro and S. Rana (eds.), *Advanced Materials for Defense*, 4, Springer Proceedings in Materials, https://doi.org/10.1007/978-3-030-34123-7_10

structural stiffness tend to manifest high dynamic back face deflection during ballistic impact, which can lead to a higher risk on injury in case of non-perforating impacts, also known as Behind Helmet Blunt Trauma (BHBT). These injuries can include bruising, skull fractures, concussion and traumatic brain injury, possibly lethal. In order to assess this risk on BHBT, different experimental methodologies have been developed, of which the most common is the use of a witness backing material (typically clay) to measure the back face signature after impact. It is, however, not an easy task to link these measurements to the real risk on trauma. Nevertheless, due to the recent development of reliable finite element models of the armour system and the human body, these technologies can be used to try to establish the link between the experimental results from evaluation trials using a witness material, and the actual risk of injury.

2 Numerical Setup

The use of numerical simulation is a fundamental tool in the design and development of armour and protection systems. As previously stated, the main objective of this work is to compare the two different armours representative of an advanced ballistic helmet design and to predict the associated trauma from bullet impacts on these targets.

The threat consists of a 9 × 19 mm lead-cored and brass-jacketed bullet at 3 different velocities in the normal to the helmet shell: 400, 430 and 450 m/s. The total mass of the projectile is 8 g and the geometry of the bullet was based on a drawing provided by the FN industry (Belgium). The constitutive behaviour of brass is described using the Johnson-Cook [1] strength and failure models and the lead core is described by the plastic-kinematic material models, adequate for isotropic and kinematic hardening plasticity with the option of including rate effects, where the yield stress is scaled using the Cowper-Symonds model [1]. The projectile was meshed with 17,000 solid elements and the contact between the core and the jacket is eroding surface-to-surface without adhesion between the two parts. The projectile was discretised using hexahedral constant stress solid elements, as can be seen

Fig. 1 Final element model of the 9 mm bullet considered for this application

Table 1 Material properties of the components of the 9 mm bullet [1, 2]

Johnson Cook material model for the brass jacket						
ρ (kg/m^3)	G (GPa)	A (GPa)	B (GPa)	C	n	m
8800	44	0.112	0.505	0.009	0.42	1.68
D1	D2	D3	D4	D5		
0.54	4.89	−3.03	0.014	1.12		
Plastic kinematic material model for the lead core						
ρ (kg/m^3)	E (GPa)	ν	E_{tan} (GPa)	β	C (/s)	P
11,270	17	0.4	0.015	0.2	600	3

in Fig. 1. The corresponding material properties of the brass and lead composing respectively the jacket and the core are summarised in Table 1.

Two different armours representative of a modular ballistic helmet design are considered in this study, made of 17 layers of Kevlar® 29 (0.5 mm thickness for each layer with total thickness of 8.5 mm) [3] and 53 layers of Dyneema® HB80 (each layer 0.148 mm with thickness of 7.8 mm) [4], respectively. In both models each ply was modelled with 1 element in the through-thickness direction. The two helmet materials and the bullet are modelled with solid elements. In order to reduce the computational cost, the helmet was simulated as a 10 × 10 (cm^2) flat plate with fixed edges to avoid the complex geometry of the helmet shell without a retention system, and with a fixed 7 mm standoff distance.

The two composite materials are modelled with solid elements. The material model used to simulate Kevlar® 29 is an elasto-plastic material, for which the strength values are taken to form an ellipsoidal surface with the axes parallel to the axes of orthotropy. The material model used to simulate Dyneema® HB80 is an orthotropic material with optional simplified damage and optional failure for composites.

The constitutive behaviour of Kevlar® 29 with epoxy is described by the damage model of Chang and Chang [1], implemented in the LS-Dyna® software with the composite failure material model. The constitutive numerical material model used for Dyneema® HB80 is the orthotropic simplified damage [1]. The material properties for Kevlar® 29 with epoxy and for Dyneema® HB80 were obtained from the literature [4, 5].

For both protections, each layer was considered to be independent from each other in order to properly apply the contact algorithm automatic surface-to-surface tiebreak with two failure criteria, based on thresholds of normal and the shear failure stresses. Both protection are fixed on the edges and the head is fixed by the base of the neck.

The head dummy model used in this article is provided by Livemore Software Technology Corporation (LSTC). This model is based on the Hybrid III crash test dummy as used in the automotive industry [6]. LSTC provides the complete body model but for this research only the head and neck were required. Figure 2a shows the numerical model of the head with the composite plate (Kevlar® 29 with epoxy or

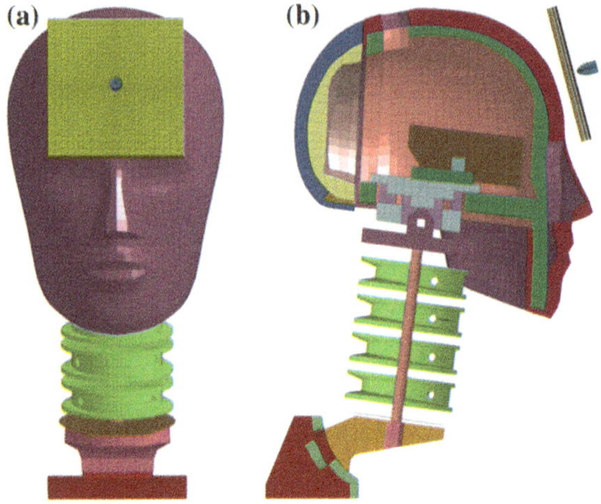

Fig. 2 Head model: **a** frontal view of the numerical head with the composite plate with the 9 mm projectile, **b** side view of the numerical head with the plate and the projectile

Dyneema® HB80) with the 9 mm projectile. This head and neck version contains the Hybrid III 50th Percentile Dummy and Fig. 2b shows the section of the numerical head with the plate and the projectile.

3 Injuries

A number of head impact criteria are available to estimate the injury levels, which can be separated in three main groups: (i) those based on linear accelerations measured at the centre of gravity of the head; (ii) those based on linear and angular (rotational) accelerations at the centre of gravity of the head, and (iii) those derived from measured or estimated brain stress and strain levels. This work focuses on a criterion from the first group, the Head Injury Criterion (HIC).

The HIC method is the most widely used in the automotive industry. The biggest disadvantage of this injury method is the fact that this method does not take into account injuries resulting from rotation induced accelerations. Additionally, it does not allow for an objective identification of the type of injury [7]. The HIC is a numerical scalar factor value that can be obtained by integrating the acceleration history on a limited time interval, using the following relation [8]:

$$HIC = \max\left\{(t_2 - t_1)\left[\frac{1}{t_2 - t_1}\int_{t_1}^{t_2} a(t)dt\right]^{\alpha}\right\}$$

where $a(t)$ is the resultant linear acceleration time history, measured in multiples of g, $\alpha = 2.5$ and t_1 and t_2 are two times instants chosen in such a way as to maximise the HIC result, subject to the constraint that they do not differ by more than a prescribed interval. The prescribed time limit is 15 ms (know as HIC_{15} criterion) to allow for the analysis of shorter duration impacts, such as those resulting from terminal ballistic phenomena [8].

Once the HIC value is known, the assessment of injuries can be done using the Abbreviated Injury Scale, where the probability of a specific level of head injury can be estimated for a specific HIC value [9].

In general, the impact responses were described in terms of head acceleration and impact force and therefore depend on the inertial properties of the head and the surface impacted. During direct head impacts, skull deformation can lead to a number of fracture patterns depending on the impact conditions. The three main types of fracture include penetrating fractures, depressed comminute fractures and remote linear fractures [10]. The risk on skull fracture from a ballistic impact is mainly due to a direct loading path from the projectile. The risk on skull fracture is determined by the back face signature (BFS) of the helmet shell during an impact event. Analyzing the available literature for the maximum allowable dynamic loading of the head (generally based on experiments on cadavers) a threshold in the range of 4.0–6.2 kN seems to have been identified by several researchers for frontal head impact. In these studies, mainly drop tests against a rigid flat surface were performed [11].

4 Results and Conclusions

Tests were done for two different composites materials and for three different impact velocities. These velocities were chosen to ensure non-perforation of the helmet for both composite materials (poly-aramid and poly-ethylene). To determine these velocities, a parametric study was carried out to determine the value of the ballistic limit.

Figures 3 and 4 show the evolution of the contact force between the plate and the skin and the Head Injury Criterion (HIC), respectively, as a function of the impact velocity for the two composite materials. In the case of Kevlar® 29 the contact force increases steeply, when compared with the Dyneema® HB80, where the forces are almost zero. The HIC values for Dyneema® HB80 are also shown to be much lower than for Kevlar 29. The HIC value for Kevlar® 29 is 1068 when the initial velocity is 450 m/s and corresponds to level 3 injuries (see Table 1) in the Abbreviated Injury Scale (AIS). Nevertheless, in all cases the contact force is under the minimum force found in literature, which is 4 kN (Table 2).

Although the maximum contact force criterion for skull fracture does not indicate any severe risk on skull fracture for both helmet configurations, the Head Injury Criterion (HIC) shows significant risk on injury for higher impact velocities in the case of the poly-aramid helmet shell configuration. Contrary to 'common belief', the poly-ethylene fibre material shows less risk on BHBT than the poly-aramid fibre material. On the contrary: they seem to reduce the risk compared to the legacy poly-aramid solutions. A probable explanation can be related with the number of

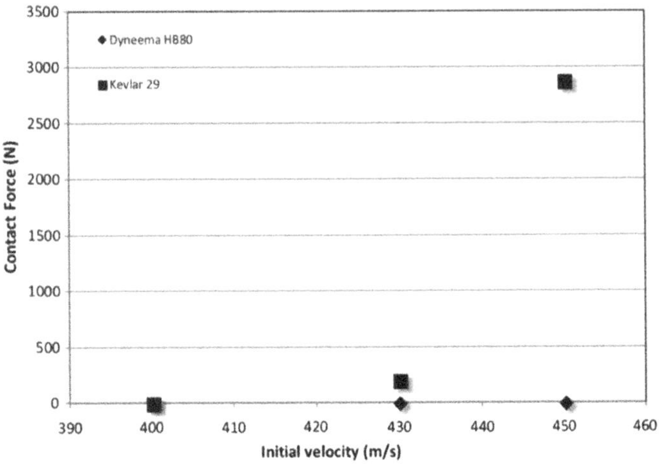

Fig. 3 Contact force as a function of initial velocity for the two composite materials

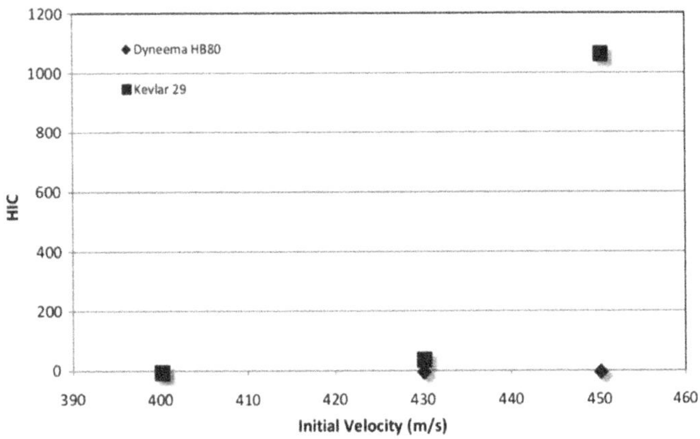

Fig. 4 HIC value as a function of initial velocity for the two different materials

Table 2 The HIC, AIS codification by severity

HIC (15 ms)	AIS code	Level of brain concussion and head injury
135–519	1	Headache or dizziness
520–899	2	Unconscious less than 1 h—linear fracture
900–1254	3	Unconscious 1–6 h—depressed fracture
1255–1574	4	Unconscious 6–24 h—open fracture
1575–1859	5	Unconscious longer than 24 h—large hematoma
>1860	6	Non survivable

different layers for both helmets. The poly-ethylene fibre material has three times more layers than the poly-aramid solution, and this can lead to an increase of the energy absorption.

For BHBT, this is quite straightforward using an instrumented headform.

References

1. Hallquist J (2012) LS-Dyna keyword user's manual, vol II. Livermore Software Technology Corporation (LSTC)
2. Krishnan K, Rajan SD, Belegundu AD (2008) A general optimization methodology for ballistic panel design. Paper presented at the EngOpt 2008 conference, Rio de Janeiro, Brasil
3. van Hoof J, Cronin DS, Worswick MJ, Williams KV, Nandlall D (2001) Numerical head and composite helmet models to predict blunt trauma. Paper presented at the 19th international symposium on ballistics, Interlaken, Switzerland
4. Anctil B, Bayne T, Williams K, Fourny P, Craigie I, Savard A (2014) Compression resistance testing of combat helmets and the effects on ballistic performance. Paper presented at the Personal armour safety symposium (PASS), Cambridge, UK
5. Adams B (2003) Simulation of ballistic impacts on armored civil vehicles. Eindhoven University, Eindhoven
6. LSTC dummy and barrier models for LS-DYNA (2017). Available from: http://www.lstc.com/download/dummy_and_barrier_models
7. Deck C, Willinger R (2008) Improved head injury criteria based on head FE model. Int J Crashworthiness 13(6):667–678
8. Henn HW (1998) Crash Tests and the Head Injury Criterion. Teach Math Appl Int J IMA 17(4):162–170
9. Begley MR, Zok FW (2013) Optimal material properties for mitigating brain injury during head impact. J Appl Mech 81(3):031014-031014–5
10. Raymond DE (2008) Biomechanics of blunt ballistic temporo-parietal head impact. Wayne State University, Detroit, MI
11. Schmitt K-U, Niederer PF, Cronin DS, Muser MH, Walz F (2014) Trauma biomechanics: an introduction to injury biomechanics, 4th edn. Springer, Berlin, Heidelberg

Vibration Analysis of High Pressure Composite Carbon Fiber Vessels Subjected to Mechanical Impact

Mirosław Bocian, Maciej Panek, Dariusz Pyka and Noemie Alexandre

Abstract Composite high-pressure vessels have been used in many engineering fields. Contrary to the civil sector, where the quantity of impact incidents is relatively low, in the military the application of composite cylinders under circumstances that lead to mechanical impacts occurs far more frequently. Since the standard tests are time and cost consuming, other methods to monitor the operational status of COPV have been searched for. One of them could be the vibration-based method. In this work, three groups of composite overwrapped pressure vessels were examined. In the first group they were non-impacted, while in the second and third ones they were subjected to two different load levels. The modal test and FEM simulations were performed to investigate possibility of damage detection based on the vibration analysis. It was proved that damage induced by the impact load has a noticeable influence on natural frequencies, and thus the modal analysis seems a useful tool for the impact detection in carbon epoxy overwrapped pressure vessels.

1 Introduction

Composite overwrapped pressure vessels (COPV) have been extensively used in many engineering fields. COPVs can be used as energy containers for vehicles as well as power generators and generally for different gases storage. Despite the low energy

M. Bocian · M. Panek (✉) · D. Pyka
Wroclaw University of Science and Technology, Wroclaw, Poland
e-mail: maciej.panek@pwr.edu.pl

M. Bocian
e-mail: miroslaw.bocian@pwr.edu.pl

D. Pyka
e-mail: dariusz.pyka@pwr.edu.pl

N. Alexandre
CEA, Le Ripault, Monts, France
e-mail: noemie.alexandre@cea.fr

R. Fangueiro and S. Rana (eds.), *Advanced Materials for Defense*, 4, Springer Proceedings in Materials, https://doi.org/10.1007/978-3-030-34123-7_11

density [13] this kind of propulsion can be successfully applied in, for example, stealth vehicles and quiet generators especially desirable on the battlefield.

Notwithstanding the fact that the quantity of vehicles accidents is relatively low in the civil sector, storage cylinders are exposed to damage occurring from mechanical impacts such as a drop of a cylinder, a drop of objects, etc. In military vehicles which can move through rough terrain, on the battlefield and in many difficult conditions, the probability that they will be subjected to a broad range of mechanical impacts, either usual or accidental, is far higher.

Damage mechanisms occurring in such composites include fiber breaks, matrix cracking and delamination, which may not necessarily be visible from the outside. Thus visual inspection could not give reliable information about the severity of damage during periodical inspections or when an impact is expected. An impact may lead to significant reduction in the residual strength of COPV [1, 2, 8, 14], therefore knowledge about how severe the impact was could be essential for determination whether the equipment can safely stay in service or not.

Since the standard tests, such as pressure test or acoustic emission test (combined with pressurization), are time and cost consuming other methods to monitor the operational status of COPV have been searched for. One of them could be the vibration-based method [4, 9, 11, 15]. The modal analysis method gives information about modal parameters of the object under examination such as natural frequencies, mode shapes and modal damping which depends on mechanical properties and the shape of the structure. Despite the scale of the phenomenon, the changes in the examined object, for example as a consequence of the impact damage, may cause that modal parameters will change [5, 6, 9, 11].

The modal analysis of the equipment is relatively inexpensive and does not require vessel pressurization therefore it seems to be a cheap, quick and safe way for the vessel condition assessment.

2 Storage Vessels Examinations

At present, the type IV COPV is widely used for high-pressure compressed gas storage (e.g. hydrogen). In this study, the 36 l carbon fiber composite high-pressure vessels were used in examinations. The test was made for 6 vessels divided into 3 groups, each consisting of 2 vessels. The first one was subjected to 1 kJ impact (0.6 kJ absorbed), in the second the impact energy was equal to 3 kJ (z1 and z2 absorbed 1.2kJ and 1.5 kJ respectively) and the third included non-impacted cylinders.

The impact test was performed using a drop tower with the hemispherical impactor tip shape. The impact occurred in the middle of the cylindrical part (Fig. 1), perpendicularly to the vessel surface. The impacts were carried out at the service pressure of 70 MPa.

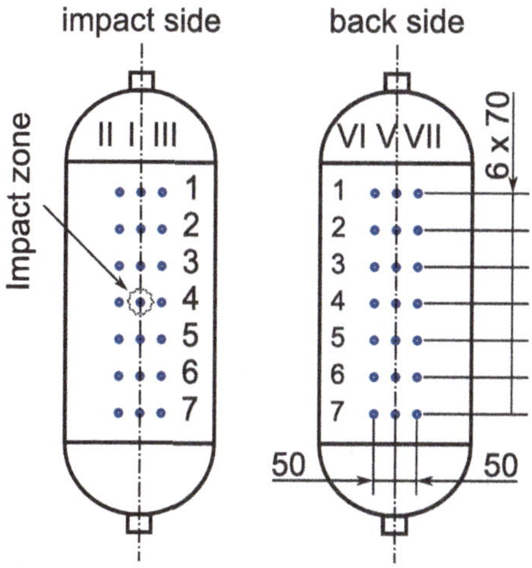

Fig. 1 Positioning of measurement points for impacted vessels

2.1 *Modal Test*

Measurements were performed on empty vessels suspended on a rope in the vertical orientation, which enabled their free vibrations. To avoid the change in mass distribution when the accelerometer was moving, which would result in natural frequency change as well as the necessity to unmount and remount the accelerometer at the different measurement points, the roving hammer method was chosen. The authors decided not to perform the test at the whole vessel but only in a few specific chosen lines (along the cylinder) to simulate the service conditions in the situation when the place of impact is not necessarily known. The measurement points were located on the cylindrical part of the vessel. 6 parallel lines were designated (only three in the case of the not impacted vessel), each of them consisted of 7 evenly spaced points. The line I always ran through the center of the impact zone while II and III were located on the sides, the other three were on the opposite side (see Fig. 1). Point no. 4 in each line was positioned at the center of the vessel.

Vibrations in COPVs were excited with the small modal hammer (PCB, model 086C03) and measured with light weight (1.7 g) acceleration sensor (PCB, model A353B17). It was to prevent the induction of too big force on the composite (possible surface damage), and in case of the accelerometer, the small weight affected modal parameters to a very small extent. Signals from the sensors were collected and processed by the dynamic signal analyzer (HP, 35665A) with the appropriate setup such as the force window, the exponential window, etc. The measurements were made separately in each line. Excitation was induced three times for each point 1–7 along the line. After finishing measurement in one line, the acceleration sensor was moved to the position above the point 4 in the next line, and later similarly. The acceleration sensor was attached perpendicularly to the vessel surface with the use of cyanoacrylate adhesive mounting. For each measurement point the frequency response function (FRF) in terms of acceleration was calculated - $A(\omega)$ (*accelerance*).

2.2 Numerical Model

Since the modal analysis was conducted separately for each line therefore there was no possibility of obtaining mode shapes for vessels as the whole structure. Because of that it was decided to prepare a simplified numerical model and compare experiment results with results of FEM simulations. The numerical modal analysis was performed in ABAQUS. The vessel was modeled as a solid element consisting of three parts: an aluminum boss section, a carbon fiber composite overwrap and a relatively thick polyethylene liner. The composite material properties were determined on a basis of homogenization method [3, 12]. In first step, the fiber percentage amount was designated, afterwards the representative RVE cell was modeled, and subsequently material properties were calculated. FEM simulations were based on the solid method with hexa elements. Two cases were considered, the first one when the vessel was intact and the second when the cylinder was damaged after the impact (3 kJ energy). In both simulations as well as in the case of the modal test the internal pressure was equal to 0 Pa. The simplified shape and dimensions of the breakage was chosen on a basis of the computer tomography scan of the impacted vessel.

3 Results and Analysis

During measurements one considerable question appeared: *does this method has rewarding repeatability in case of composite vessels?* For this reason, additional tests were conducted at the same place to repeat measurements after remounting the acceleration sensor. The measurement data analysis revealed that the repeatability was satisfactory (see Fig. 2).

Natural frequencies and the corresponding modes along the lines were determined as a result of the data processing. The analysis of natural frequencies proved that in

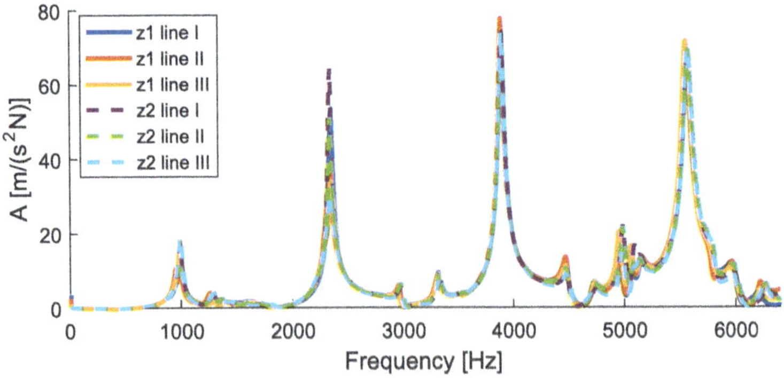

Fig. 2 Cumulative frequency response functions (FRF) obtained for two intact vessels (0 kJ)

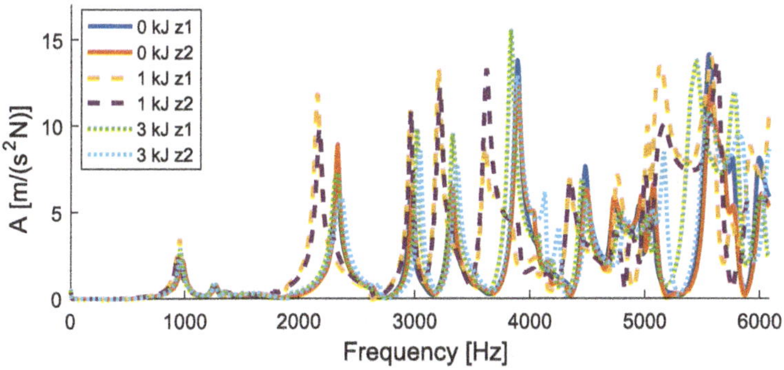

Fig. 3 Summed frequency response functions (FRF) obtained at point 5 in line 2

case of the impacted vessel the considerable frequency shift occurred for almost all peaks. Even for two first peaks the shift to lower frequencies was noticeable, especially in the line near the impact zone (Fig. 3).

The analysis of frequency graphs separately for each measurement point is difficult and may distract attention from general trends, thus the summed, for each point in line, frequency response functions (cumulative FRFs) were calculated for each measurement line. Based on the results obtained for the non-impacted vessels it can be noted that cumulative FRFs are consistent and repeatable both in case of different measurement lines at this same cylinder as well as for different vessels (Fig. 2). It testifies about high repeatability considering the technological aspect (stable product and manufacturing parameters, for example the winding and curing process or materials parameters) as well as regarding the measurement methodology including the sensor mounting and the modal hammer excitation. It can be supposed that this kind of method for comparing the transfer functions of vessels could be also used for detection of defects of inner layers in pressure vessels just after manufacturing.

Determination of modal shapes based on experimental results did not give information about spatial deformation of the whole vessel, even for the cylindrical part. There was no information for example which mode had a bending shape and which had a complex spatial shape or what kind of shape occurred in this structure. The simulation results showed that the first and the third ones were bending modes while most of the others had a complex spatial shape (Fig. 4). As in case of the non-impacted vessel, the modes shapes of the impacted vessels stayed unchanged, however natural frequencies decreased when the cylinder damage is considered. Such a behavior was expected, documented [7, 10] and generally consistent with experiment results. Unfortunately, the simulations resulted in a plenty of modal shapes which were difficult to correlate with experiment results.

The cumulative FRF graphs for the impacted vessels (1 and 3 kJ) depict that repeatability was not as rewarding as in case of the not impacted ones. The likely causes are, for the same impact energy, differences in the internal and external damages in the vessels. However, some tendencies in changes in resonant frequencies are

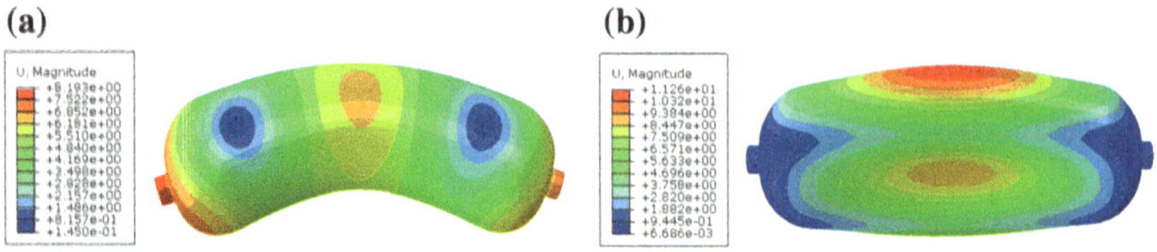

Fig. 4 Examples of modes obtained from simulation of non-impacted vessel: **a** first mode—964 Hz; **b** second mode—988 Hz

noticeable. To make it more visible the *mobility* was calculated ($M(\omega)$) and subsequently the frequency band (0–6400 Hz) was divided on sections, each 400 Hz wide. For each section separately, the absolute amplitude of frequency lines within 400 Hz range was summed and enumerated.

Exemplary mobility graphs for chosen lines are shown in Fig. 5. The amplitudes for line II which was close to the impact zone were noticeable larger. It means

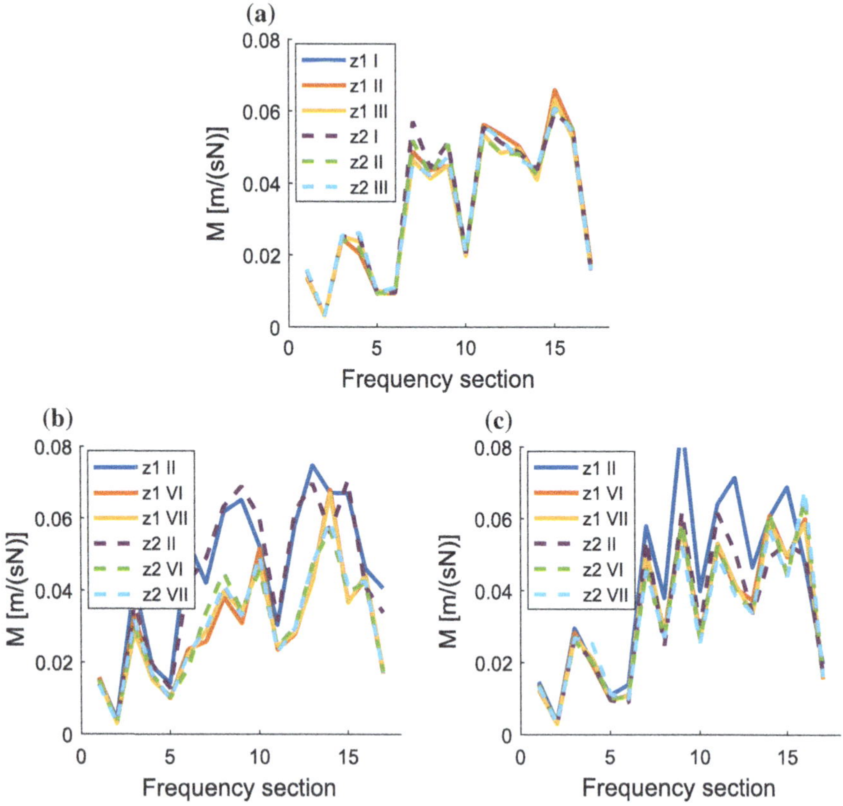

Fig. 5 Cumulative absolute frequency response functions: **a** non-impacted vessels; **b** cylinders after 1 kJ impact; **c** cylinders after 3 kJ impact

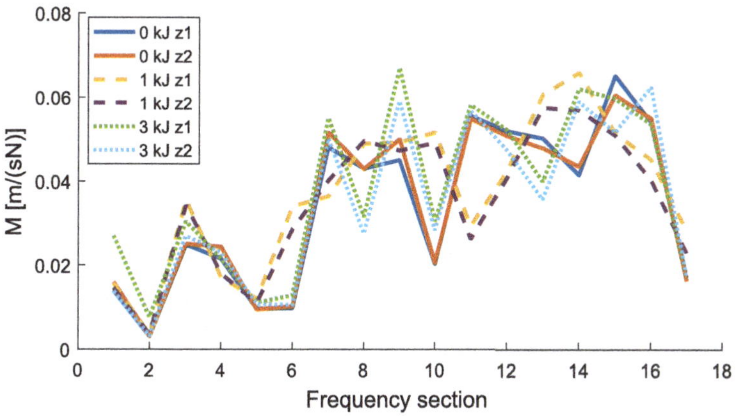

Fig. 6 Averaged, cumulative absolute FRFs calculated for examined vessels

that for the same range of force values the cylinder replied with the higher velocity amplitude. It testifies that the intact vessel which had been designed to resist external and especially internal forces strongly suppressed vibrations and when the damage occurred the damping ratio decreased locally. This effect can be useful to localize the impact zone, at least roughly.

One of method for assessing the vessel condition could be the calculation of average cumulative FRF graphs obtained for the whole vessel. It can be designated as the sum of the line cumulative mobility functions FRF divided by the quantity of measurement lines (Fig. 6). The comparison with the non-impacted cylinder is expected to show differences the occurrence of which indicates that the structure has been impacted. In such a case the damage zone could be searched or the decision made whether to reject the vessel or not.

4 Conclusions

The FEM model gave a general overview of the mode shapes in the examined COPVs, which can be helpful when experimental data does not provide such information, however it needs to be improved so that they are consistent when it comes to values of natural frequencies.

The modal analysis and simulations were conducted for the vessels under different impact conditions. The data analysis proved that the values of natural frequency changed when the vessel was impacted with different energy values. It was difficult to indicate the specific parameter which could quantify what kind of damage occurred in the vessel but some trends could be observed. The analysis results showed that greater changes in frequencies and mode shapes were visible for measurement points located close to the impact zone than for others. This phenomenon could be helpful for localizing impact zones which are invisible during service as well as for product quality control.

In this study the authors demonstrate that the modal analysis is a useful tool for the impact detection in carbon epoxy overwrapped pressure vessels, but further work is required to describe the interdependence between the impact energy and modal parameters.

Acknowledgements The research leading to these results has received funding from the European Union's Seventh Framework Programme (FP7/2007–2013) for the Fuel Cells and Hydrogen Joint Technology Initiative under grant agreement no. 621194, http://www.hypactor.eu/.

The research was also supported in part with computational resources at Wroclaw Centre for Networking and Supercomputing, grant agreement no. 45, http://www.wcss.pl/.

References

1. Ballére L, Viot P, Lataillade JL, Guillaumat L, Cloutet S (2009) Damage tolerance of impacted curved panels. Int J Impact Eng 36:243–253
2. Blanc-Vannet P (2017) Burst pressure reduction of various thermoset composite pressure vessels after impact on the cylindrical part. Compos Struct 160:706–711
3. Czapliński T, Maciejewski Ł, Zietck G (2015) Modeling of high pressure composite vessels. In: ICCST/10th International conference on composite science and technology
4. Dessi D, Camerlengo G (2015) Damage identification techniques via modal curvature analysis: overview and comparison. Mech Syst Signal Process 52:181–205
5. Ewins DJ (2000) Modal testing : theory, practice, and application. Wiley-Blackwell
6. Heylen W, Sas P (2006) Modal analysis theory and testing. Katholieke Universteit Leuven, Departement Werktuigkunde
7. Krawczuk M, Ostachowicz W, Zak A (1997) Modal analysis of cracked, unidirectional composite beam. Compos Part B Eng 28(5–6):641–650
8. Lasn K (2016) The sensitivity of the burst performance of impact damaged pressure vessels to material strength properties. In: IOP conference series: materials science and engineering, vol 139
9. Manoach E, Warminski J, Kloda L, Teter A (2017) Numerical and experimental studies on vibration based methods for detection of damage in composite beams. Compos Struct 170:26–39
10. Mehdi H, Upadhyay R, Mehra R, Singhal A (2014) Modal analysis of composite beam reinforced by aluminium-synthetic fibers with and without multiple cracks using ansys. Int J Mech Eng (IJME) 4:70–80
11. Perez MA, Gil L, Oller S (2014) Impact damage identification in composite laminates using vibration testing. Compos Struct 108:267–276
12. Pierard O, Friebel C, Doghri I (2004) Mean-field homogenization of multi-phase thermo-elastic composites: a general framework and its validation. Compos Sci Technol 64(10–11):1587–1603
13. Pointon KD, Lakeman JB (2007) Prospects for hydrogen as a military fuel. In: Assessment of hydrogen energy for sustainable development. Springer, pp 97–106
14. Weber M, Devilliers C, Guillaud N, Dau F, Gentilleau B, Nony F, Villalonga S, Halm D (2014) Damage tolerance of compressed gaseous hydrogen composites vessels. ECCM Sevilla
15. Zhou W, Wu Z, Mevel L (2010) Vibration-based damage detection to the composite tank filled with fluid. Struct Health Monit 9(5):433–445

Numerical and Experimental Analysis of Aramid Composites Against Ballistic Impact

M. Rodríguez-Millán, I. Rubio-Díaz, J. Aranda-Ruiz, M. M. Moure, J. A. Loya and M. H. Miguélez

Abstract In this work, an experimental and numerical study of the mechanical behaviour of combat helmets against ballistic impact of spherical projectiles of 1.7 g has been carried out. A numerical model, that has been calibrated and validated with experimental results of plates with different thicknesses, has been developed. Once the model has been validated, experimental tests have been carried out on the combat helmet. To conclude, it is noteworthy that the location of the impact on the helmet has a significant influence on the ballistic limit. It should be noted that the numerical model includes each layer of the combat helmet which may be beneficial for future research on helmets by using aramid or other materials.

Keywords Combat helmet · Impact · Numerical model · Experimental tests

1 Introduction

The manufacture of combat helmets involves efforts to reduce their weight without losing impact resistance performance. This is the main objective of the military industry. A great campaign of impact tests is required for the development of a combat helmet. The complexity of experimental testing justifies the use of numerical models that can predict the mechanical behaviour of personal protection systems against ballistic impact. Using these numerical tools, the helmet design can be optimized for different ballistic threats.

This work has focused on the analysis of the V50 velocity using small projectiles of spherical geometry. A combined experimental and numerical methodology is used

M. Rodríguez-Millán (✉) · I. Rubio-Díaz · M. M. Moure · J. A. Loya · M. H. Miguélez
Department of Mechanical Engineering, University Carlos III of Madrid,
Avenida de la Universidad 30, Leganés, Madrid, Spain
e-mail: mrmillan@ing.uc3m.es

J. Aranda-Ruiz
Department of Continuum Mechanics and Structural Analysis, University Carlos III of Madrid,
Avenida de la Universidad 30, Leganés, Madrid, Spain

R. Fangueiro and S. Rana (eds.), *Advanced Materials for Defense*, 4, Springer
Proceedings in Materials, https://doi.org/10.1007/978-3-030-34123-7_12

to achieve this objective. Few studies in the literature have combined experimental tests and numerical simulations [1–3]. Tham et al. [3] conducted experimental tests and numerical ballistic impact simulations on a Kevlar helmet using 11.9 g spherical projectiles launched up to 205 m/s. They completed the study by numerically analyzing the mechanical behaviour of the helmet against FSP and FMJ projectiles. Tan et al. [2] carried out experimental tests and numerical simulations of frontal and lateral impact with a spherical projectile of 11.9 g launched at 220 m/s. Recently, Rodríguez-Millán et al. [1] analysed experimental and numerically the mechanical behaviour of the combat helmet developed by a Spanish company according to the NIJ 0106.01 [4] and STANAG 2920 [5] standards, obtaining good predictions.

Other researchers have focused on the numerical analysis of the combat helmet without experimental tests due to the difficulty of performing them [6–9]. Van Hoof et al. [6] numerically assessed the mechanical response to ballistic impact of aramid helmets and plates. They observed that higher deformation occurred in the helmet than in the plate. Helmet simulations indicated that the damage caused by the impact is very localized and its effects are limited to the impacted area. The overall movement of the helmet is negligible compared to the plate. Lee and Gong [7] developed numerical simulations to evaluate the effectiveness of the combat helmet considering different internal damping systems. Recently, Li et al. [8] developed a numerical model to simulate the ballistic behaviour of the Advanced Combat Helmet (ACH) to analyse different variables: helmet size, angle of impact and impact location. They concluded that at the same impact velocity, the small helmet has more significant deformation and that the worst location for impact was the frontal one. Palta et al. [9] carried out numerical simulations on the Advanced Combat Helmet (ACH) using 1.1 g FSP, 9 g FMJ and .223 rifle projectiles. They found that while the ACH is valid against 9 mm bull et impacts, it is unsuitable for the impact of .223 rifle ammunition.

In this work, spherical projectiles have been chosen because around 20% of the wounds found in soldiers and produced by an improvised explosive device (IED) were in the form of a spherical fragment, even reaching up to 70% in the case of mines or RPGs [10].

The objective of this work is the analysis of the effect of parameters such as the geometry of the target (helmet and plate) and the impact velocity on the penetration mechanisms. As a second contribution, a numerical finite element model is developed to predict the response of a combat helmet under ballistic impact. A novelty in the numerical model of the combat helmet is that each layer of aramid has been discretized with an element through the thickness. The combat helmet consists of more than twenty layers along the thickness. Generally, it has been modelled without considering one element per layer due to the high computational cost [1, 2, 9, 11–13].

2 Experimental Methodology

This section describes the methodology carried out in the experimental tests for the calibration of the numerical model.

2.1 Material

Material of the used plates and combat helmet is formed by resin preimpregnated aramid sheets and thermoformed to acquire rigidity. The main characteristics of the aramid composite are its resistance, excellent energy absorption capacity and good strength-to-weight ratio, which makes it a very attractive material for the defence and transportation sectors.

For the ballistic impact tests on $100 \times 100\,\text{mm}^2$ plates, two different areal densities of 4.43 and $8.86\,\text{kg/m}^2$ have been taken into account. The tests in the combat helmet have been carried out considering an areal density of $8.86\,\text{kg/m}^2$.

2.2 Experimental Device

The experimental tests were carried out with a 7.62 mm caliber pneumatic gun that is capable of launching projectiles up to 6 bar pressure. The projectiles used were steel spheres of 1.7 g and 7.5 mm of diameter. The range of impact velocities used was $350\,\text{m/s} < V_0 < 850\,\text{m/s}$. In all tests, the projectiles did not suffered any deformation, damage or erosion after impact. In the case of the plate tests, the specimens were supported on their contour and at two points to prevent slipping during impact. However, the combat helmet is supported on its side by a swiveling tool (see Fig. 1) that allows impacts to be made in 5 different locations (front, sides, back and top). A Photron FastCam SA-Z high-speed digital camera has been used to record the impact sequence to obtain both the impact speed and the residual speed in the case of full penetration.

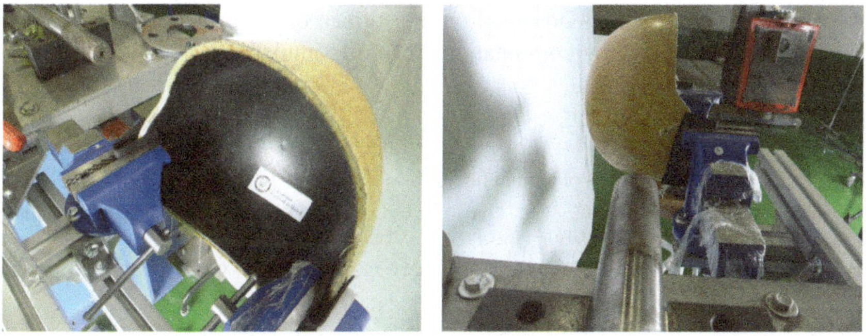

Fig. 1 Experimental setup

3 Development of the Numerical Model

This section describes the finite element model used for the analysis of the mechanical behaviour of aramid composites under ballistic impact. The model has been developed in the ABAQUS/Explicit finite element commercial code using the lagrangian method, being validated with experimental data. Three solids have been modeled: the projectile, the fixing structure and the target (plate or helmet). The geometry of the target and projectiles is the same as that used in experimental tests.

3.1 Constitutive Model for Aramid Composite

The ACH combat helmet is made of aramid fibre composite material, whose mechanical properties depend on the direction of the fibres. The behaviour of the material has been considered linear elastic up to failure, whose mechanical properties are shown in Table 1 [7].

It is important to take into account the different failure mechanisms that can occur in this type of material: interlaminar failure and intralaminar failure. To take these mechanisms into account, the Hashin failure criterion has been implemented through a VUMAT user subroutine. The values of the parameters used in the Hashin criterion (see Table 2) have been obtained from the scientific literature [7].

3.2 Definition of Numerical Model

The projectile is defined as a rigid, non-deformable solid, since in experimental tests no plastic deformation was observed on the surface of the projectile after impact. This definition reduces the computational cost of simulations. The projectile was modeled in AISI 4340 steel.

Table 1 Mechanical properties of aramid fiber [7]

E_1 (GPa)	E_2 (GPa)	E_3 (GPa)	G_{12} (GPa)	G_{13} (GPa)	G_{23} (GPa)	ν_{12} (−)	ν_{13} (−)	ν_{23} (−)	ρ (kg/m^3)
18.0	18.0	4.5	0.77	2.6	2.6	0.25	0.33	0.33	1230

Table 2 Failure properties of aramid fiber composite [7]

X_{1t} (MPa)	X_{1c} (MPa)	X_{2t} (MPa)	X_{2c} (MPa)	X_{3t} (MPa)	X_{3c} (MPa)	S_{12} (MPa)	S_{13} (MPa)	S_{23} (MPa)
555	555	555	555	1050	1050	77	1060	1086

Regarding the plate and the helmet, each layer is meshed with C3D8R elements and with one element per layer along the thickness. Different areas of the helmet have been meshed to reproduce the frontal and upper impacts.

To model the interlaminar behaviour, a cohesive interaction between layers has been defined, based on the law of traction-separation in quadratic formulation. The maximum admissible stresses for normal (S_n) and tangential (S_s) directions are 34.5 MPa and 9.0 MPa, respectively [7]. The contact between the projectile and the plates is defined by a non-penetration restriction and a coefficient of friction [14]. The *hard contact* option allows you to automatically adjust the stiffness, generated by the *penalty contact algorithm*, to minimize penetration without negatively affecting the increase in time. For friction effects, a dynamic coefficient of friction μ is assumed to be 0.33 between steel and composite.

3.3 Validation of Numerical Model with Experimental Tests

The validation of the numerical model was carried out with the impact tests on the plates and was evaluated in terms of residual velocity. The use of a mesh density of one element per layer thickness was sufficient to accurately predict the impact response of the material. A comparison between experimental and numerical results regarding residual velocity (V_r) versus impact velocity (V_0) was made for both areal densities, as can be seen in Fig. 2. These results have been adjusted using the expression proposed by Recht and Ipson [15]:

$$V_r = \left(V_0^{\kappa} - V_{bl}^{\kappa}\right)^{1/\kappa} , \quad (1)$$

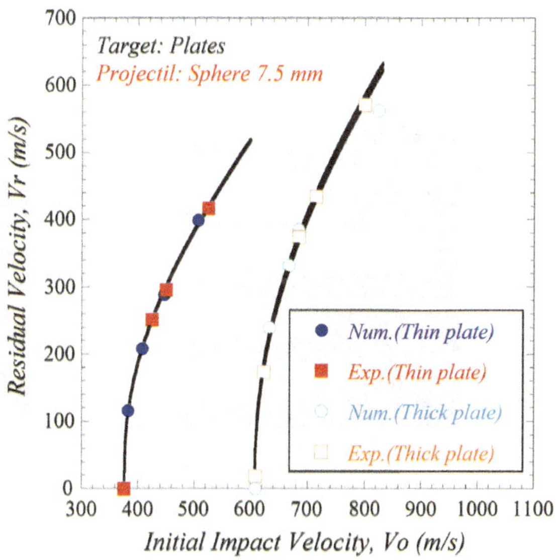

Fig. 2 Comparison between experimental and numerical results for impact tests on aramid plates with different areal density

where κ is an adjustment parameter that depends on the shape of the projectile and V_{bl} is the value of the ballistic limit. The adjustment parameters obtained for thick plates are $\kappa = 2.30$ for experimental data and $\kappa = 2.36$ for numerical predictions, obtaining an error of 2%. For the case of thin plates, the values of the adjustment parameter are $\kappa = 2.52$ and $\kappa = 2.51$ for experimental data and numerical predictions respectively, obtaining an error of 1%. It can be seen that the values of this adjustment parameter are slightly higher for the thin plates. This indicates that the adjustment parameter depends not only on the projectile, but also on the target thickness.

The model provides a good correlation between the numerically results and the experimental data. Once the model has been validated in terms of residual velocity, it is used to analyse the ballistic behaviour of the combat helmet.

4 Results

Experimental tests are carried out in two combat helmets and a total of 10 impacts (two per location) in a range of impact velocities from 616 to 679 m/s. All tests are carried out above the ballistic limit of the material to obtain residual velocities. Figure 3 shows the experimental results ($V_0 - V_r$) according to the location of the impact.

Ballistic impact simulations were performed in the frontal and upper area of the helmet to analyze the influence of the impact location on the ballistic limit as well as on the value of the adjustment parameter, κ, according to the model of Recth and Ipson (Eq. 1) (see Fig. 4). The ballistic limits obtained are 533 m/s and 608 m/s for the frontal and upper impact, respectively. The adjustment parameters are $\kappa = 2.92$ and $\kappa = 2.11$ for the frontal and superior impact. The impact zone has a significant influence on the ballistic limit. Numerical simulations revealed differences of 14%

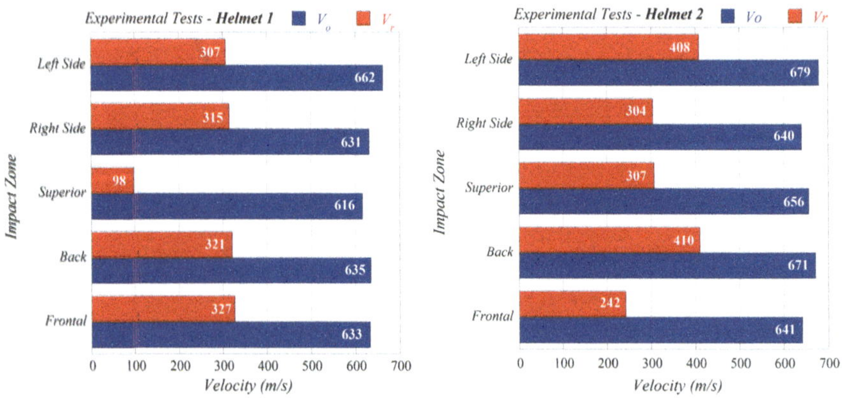

(a) Helmet 1 (b) Helmet 2

Fig. 3 Experimental results of impact on combat helmets

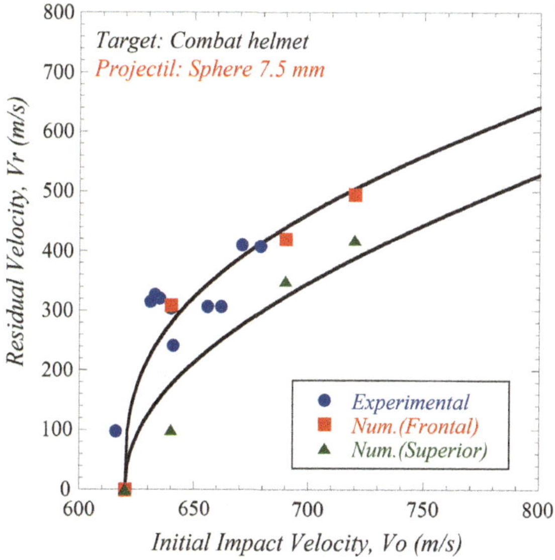

Fig. 4 Comparison between experimental and numerical results for impact tests on combat helmet

(70 m/s) depending on the impact zone. Higher impact velocity values are required for impacts on the top of the helmet; this may be due to contour conditions as well as to the geometry of the helmet at each location. It should be noted that the frontal impact is produced at approximately 50 mm from the free edge of the helmet.

The impact velocity is a parameter that has been shown to be significant. For low impact velocities, the location of the impact is not of great importance; however, at impact velocities above the ballistic limit, localization is relevant. In addition, at velocities close to the ballistic limit, a greater transfer of kinetic energy from the projectile to the helmet is produced, resulting in greater overall deformations (delamination). However, at impact velocities far above the ballistic limit, the effects are more local than global.

5 Conclusions

In this study, a ballistic analysis was carried out using numerical simulations and experimental tests to evaluate the behaviour of combat helmets with spherical steel projectiles of 1.7 g. The numerical model has been calibrated and validated with experimental tests on thin and thick plates in terms of residual velocity. It is also worth noting that the model is designed in such a way that it hast the same number of layers as in the real model. Once the numerical model has been validated, experimental tests and numerical simulations have been carried out in real combat helmets. The results revealed that the location of the impact can be influential when determining the ballistic limit. Therefore, the numerical model of combat helmet developed in this study, is capable of reproducing the mechanical response of the helmet against the impacts of spherical projectiles.

Acknowledgements This work has been carried out within the framework of the research projects DPI2017-88166-R and RTC-2015-3887-8 of FEDER program financed by the Ministerio de Economía, Industria y Competitividad of Spain.

References

1. Rodríguez-Millán M, Ito T, Loya JA, Olmedo A, Miguélez MH (2016) Development of numerical model for ballistic resistance evaluation of combat helmet and experimental validation. Mater Des 110:391–403
2. Tan LB, Tse KM, Lee HP, Tan VBC, Lim SP (2012) Performance of an advanced combat helmet with different interior cushioning systems in ballistic impact: experiments and finite element simulations. Int J Impact Eng 50:99–112
3. Tham CY, Tan VBC, Lee HP (2008) Ballistic impact of a KEVLAR helmet: experiment and simulations. Int J Impact Eng 35:304–318
4. NIJ Standard for Ballistic Helmets. Justice NI of USA (1981)
5. STANAG 2920 PPS (Edition 2)—Ballistic test method for personal armour materials and combat clothing. NATO Standardization Agency (2003)
6. Van Hoof J, Cronin DS, Worswick MJ, Williams KV, Nandlall D (2001) Numerical head and composite helmet models to predict blunt trauma. In: 19th international symposium on ballistics. Interlaken, Switzerland, pp 7–11
7. Lee HP, Gong SW (2010) Finite element analysis for the evaluation of protective functions of helmets against ballistic impact. Comput Methods Biomech Biomed Eng 13:537–550
8. Li YQ, Li XG, Gao XL (2015) Modeling of advanced combat helmet under ballistic impact. J Appl Mech 82:1–9
9. Palta E, Fnag H, Weggel DC (2018) Finite element analysis of the Advanced Combat Helmet under various ballistic impacts. Int J Impact Eng 112:125–143
10. Barrett J (2015) Design validation of future ballistic neck protection through the development of novel injury models. Ph.D. thesis. University of Birmingham
11. Li XG, Gao XL, Kleiven S (2016) Behind helmet blunt trauma induced by ballistic impact: a computational model. Int J Impact Eng 91:56–67
12. Rodríguez-Millán M, Tan LB, Tse KM, Lee HP, Miguélez MH (2016) Effect of full helmet systems on human head responses under blast loading. Mater Des 117:58–71
13. Tan LB, Chew FS, Tse KM, Chye Tan VB, Lee HP (2014) Impact of complex blast waves on the human head: a computational study. Int J Numer Methods Biomed Eng 30:1476–1505
14. Systemes D. Abaqus v6.14 Documentation—ABAQUS Analysis User's Manual ABAQUS. 2014th ed (2014)
15. Recht RF, Ipson TW (1963) Ballistic perforation dynamics. J Appl Mech 30:384–390

Comparison of Numerical Modelling Methods of Innovative Materials for Ballistic Shields

Adrianna Mackiewicz, Dariusz Pyka, Joanna Pach, Krzysztof Jamroziak and Mirosław Bocian

Abstract The article contains a comparison of selected numerical methods describing the structure of polymer composite. This work focuses on different FEM modeling methods using homogenization of composite material parameters. The estimation is based on comparing visual results of real ballistic test and numerical simulations. The main goal of this comparison is estimation of computer description accuracy of ballistic impact by 9 mm projectile type Luger.

Keywords Arabic-fibre composites · Numerical methods · Impact behavior · Ballistic shield

1 Introduction

Currently used materials for ballistic shields should have the highest possible specific strength, i.e. high strength with the lowest mass. In addition, the material should exhibit high dissipation capacity of impact load energy [1, 2]. The possibility of combining such properties is provided by polymer composites reinforced with aramid fabrics [3].

In order to estimate the dissipation of energy by a selected material in the piercing process, FEM numerical methods can be used [4, 5]. In this respect, one of the basic issues becomes the modeling of phenomena occurring during the overshooting the composite material, the structure of which consists of two basic elements, namely reinforcement and matrix. Many factors affect the behavior of these elements. This modeling requires considerable skill because it is assumed that the model should fairly accurately reflect the physical phenomenon.

A. Mackiewicz · D. Pyka · K. Jamroziak (✉) · M. Bocian
Department of Mechanics, Materials Science and Engineering, Wroclaw University of Science and Technology, Smoluchowskiego 25, 50-370 Wroclaw, Poland
e-mail: krzysztof.jamroziak@pwr.edu.pl

J. Pach
Department of Foundry, Plastics and Automation, Wroclaw University of Science and Technology, Lukasiewicza 5, 50-370 Wroclaw, Poland

R. Fangueiro and S. Rana (eds.), *Advanced Materials for Defense*, 4, Springer Proceedings in Materials, https://doi.org/10.1007/978-3-030-34123-7_13

On the basis of the literature review, there are several ways to approach the issue. One of the most popular is the modeling of the composite structure using Representative Volume Element (RVE) [6]. Next are the methods of mass points connected mutually with flexible elements [7]. Methods based on the finite element method become quite popular, where the fabric is represented by means of solid or shell elements in a direct way (discrete, three-dimensional approaches) as a mesostructure taking into account the shape and dimensions of the analyzed material [8]. The last group are mixed methods of modeling the material in direct contact with the penetrator (principal yarns), taking into account the shape and dimensions of the applied reinforcement and interlacing details, while the remaining ones as a homogenized continuous center [9].

The subject of the thesis was focused on the comparison of various numerical methods in terms of energy and quality. The research problem was considered in terms of determining the best numerical model describing a polypropylene-matrix multilayer composite reinforced with aramid fabrics.

2 Preparation of Samples and Experimental Tests

2.1 *Technology of Samples Manufacturing*

The tested laminate was produced in two stages by pressing at high temperature using a PAREN manual hydraulic press H-Type. The component materials were the aramid fabric 173 g/m^2 Twaron® 1210 dtex, basis weight and type 2200, plain weave and fabric thickness 250 μm and the Moplen HP548R polypropylene granulate with a density 0.9 g/cm^3.

The first stage involved the preparation of polypropylene films by plasticizing the granulate at 200 °C for the first 2 min without pressure, and then pressing for 2 min under load. During heating, the material plasticized and filled the mold, taking the desired shape.

The second production step included pressing the prepared polypropylene and aramid fabric layers in a steel mold. Ten layers of fabric and 11 polypropylene foils arranged alternately were used to make one ten-layer laminate sample. The materials were pre-plasticized at a constant temperature of 200 °C and then pressed under a load of 50 kN. After plasticizing and supersaturation of the fibers, the sample was allowed to cool.

2.2 *The Quasi-Static Punching Test*

The puncture test was carried out on a strength machine MTS 810 with a measuring range up to 100 kN. The test was carried out for a constant speed of the impactor, equal to 30 mm/s. The impactor load was increased until the sample was pierced while maintaining the set speed. The shape of the impactor was adopted in accordance with the projectile geometry 9 × 19 mm Luger. For each sample, 3 strokes were made at an angle of 90° with respect to the mounting plane of the sample.

2.3 *The Analysis of the Quasi-Static Puncturing Test Results*

Sample results of the conducted strength tests are presented in Fig. 1 in the form of a graph.

The analysis of the graphs shows that the matrix of the composite, which is made of a material of lower strength than fibers, is the first to be destroyed during penetrating of the composite. Disturbances in the course of dependence of force on displacement suggest stages of permanent stretching of fibers, delamination, cracking of matrix and cracking of fibers.

During quasi-static puncturing test, the tested material is in a complex stress state (positive and shearing stress). For this reason, the quasi-static puncturing test is of comparative nature. From the test results, one can be read and compared the values of the maximum forces needed to penetrate different materials. The obtained values of forces are used for further estimation of strength in ballistic tests of the tested material.

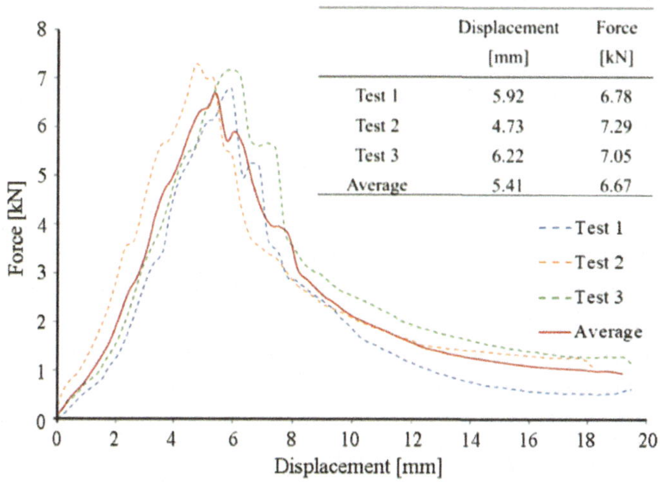

	Displacement [mm]	Force [kN]
Test 1	5.92	6.78
Test 2	4.73	7.29
Test 3	6.22	7.05
Average	5.41	6.67

Fig. 1 Empirical graphs of dependence of force on displacement for quasi-static punctures of a ten-layer composite with a polypropylene matrix reinforced with an aramid fabric

2.4 Ballistic Test

The laminate was prepared as samples 100 × 100 mm (used size was dependent on the coupling at the test standard) in order to perform ballistic track according of the European Norm CEN EN 1522. The samples were fired on the ballistic track with the use of the machine gun—9 mm Uzi and ammunition 9 × 19 mm Luger FMJ with a mass of 8 g bullets and muzzle velocity $v_{15} = 360$ m/s, which for the needs of the ballistic test has been properly prepared to meet the speed conditions specified in the standard. Velocities were measured by the Doppler radar type Weibel SL-525PE. One shot was fired for each sample, due to size limitations and maintaining a minimum distance from the edge of the sample, approximately 100 mm. During the test the following firing speed was recorded: $v_1 = 396$ m/s, $v_2 = 404$ m/s, $v_3 = 407$ m/s, $v_4 = 398$ m/s, $v_5 = 402$ m/s.

Ballistic tests were conducted to evaluate the results obtained from numerical analyzes of the punching process of the analyzed composite material. In this case, they were secondary importance, because the main goal was to develop an optimal numerical model.

3 Numerical Methods

3.1 Mathematical-Physical Models

The material penetrating phenomenon was divided into two phases. For the mathematical description of models in the elastic range (reversible), the dependence $\sigma(\varepsilon)$ was used in accordance with Hooke'a law:

$$\sigma = E\varepsilon \tag{1}$$

The above dependence is a great simplification, since the literature data [10, 11] as well as the data from quasi-static puncturing (Fig. 1) testify to the non-linearities of fiber strain in the reversible range.

The Johnson-Cook equation was used to describe the $\sigma(\varepsilon)$ dependence curve in the plastic range [12].

$$\sigma = \left(A + B\varepsilon^n\right) \cdot \left[1 + C\ln\left(\frac{\dot{\varepsilon}}{\dot{\varepsilon}_0}\right)\right] \cdot \left[1 - \left(\frac{T - T_0}{T_m - T_0}\right)^m\right] \tag{2}$$

where: ε—is the plastic strain, $\dot{\varepsilon}$—the plastic strain rate, $\dot{\varepsilon}_0$—reference strain rate, T—current temperature, T_0—room temperature, T_m—melting temperature, A, B, C, n, m—Johnson-Cook material behavior coefficients.

It applies to both plastic and elastic deformations at rate of shear greater than 10^3 s^{-1}. Johnson-Cook's designs can be used to describe virtually any material, including brittle materials, assuming that they have minimal plastic deformations before breaking. Johnson-Cook Fracture Model was used to describe the mechanism of material destruction.

$$\varepsilon^{f} = \left(D_1 + D_2 \exp D_3\sigma^{*}\right) \cdot \left[1 + D_4 \ln\left(\frac{\dot{\varepsilon}}{\dot{\varepsilon}_0}\right)\right] \cdot \left[1 - D_5\left(\frac{T - T_0}{T_m - T_0}\right)^{m}\right] \quad (3)$$

where: D_1, D_2, D_3, D_4 and D_5—input constants determined empirically, ε^{f}—plastic strain to fracture, σ *—equivalent stress.

The model of destruction is thus dependent on the spatial stress state, the deformation rate and temperature.

The substitute properties of polypropylene-saturated fibers and a layer of polypropylene-saturated fabric were determined by homogenization. For the homogenization of the fiber bundle saturated with polypropylene, the volume content of V_f fibers was assumed at 80.6%, while the total volume of reinforcement in the composite was assumed at 40% (Table 1). Assumptions were made on the basis of analysis of the sample cross-section by SEM [2, 13].

Table 1 Values of coefficients used for the mathematical description of the materials [4, 12, 14]

Symbol	Unit	Fabric			
		PP	Twaron	Fiber homogenization	Global homogenization
V_f	[–]	–	–	0.806	0.40
E	[MPa]	1.65×10^3	85×10^3	1.34×10^3	1.12×10^3
ν	[–]	0.42	0.30	0.32	0.46
The Johnson-Cook plastic strength model					
A	[MPa]	35	456	369	209
B	[MPa]	45	460	375	214
m	[–]	0.8	1.0	1.0	1.0
n	[–]	1.00	0.40	0.18	0.18
ε_0	[–]	1.0	1.0	1.0	1.0
C	[s^{-1}]	0.1	1.0	0.0046	0.0046
Model destruction Johnson-Cook					
D_1	[–]	0.068	1.40	0.068	0.068
D_2	[–]	5.382	0.28	5.382	5.382
D_3	[–]	−2.534	−0.10	−2.534	−2.534
D_4	[–]	–	–	–	–
D_5	[–]	–	–	–	–

where E Young's modulus, ν Poisson ratio

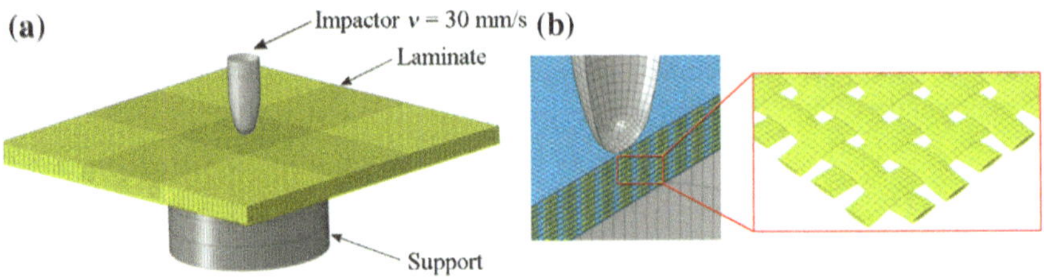

Fig. 2 Discrete systems used in quasi-static puncture simulation: **a** volumetric and layered model of laminate with global homogenization, **b** layered model with regard to fabric structure and laminate matrix

3.2 Discrete System Models

Three discrete laminate models were used to compare numerical simulations (Fig. 2). The first model of the laminate was modeled as a solid material and the material properties obtained from global homogenization were given. The second model of the laminate was modeled in the form of 10 adjacent layers with a thickness of 0.475 mm each and given the material properties obtained from global homogenization. The third model of the laminate takes into account the fabric structure with the homogenization of the properties of the supersaturated polypropylene fiber and the simplified structure of the matrix in the form of layers.

In each of the created systems, composed of a impactor, support and laminate, simplifications were applied, consisting in the non-deformation of the impactor and support model and the lack of temperature influence on the course of the laminate deformation process. The impactor and support are made of non-deformable tool steel and the puncture speed is low, which argues the use of the above simplifications.

All of the layered models of the laminate omit the interlayer adhesion forces due to the low degree of supersaturation of the fabric fibers with polypropylene. A deformable bullet model and a ten-layer laminate model with homogenization of each layer were used to simulate impact perforation and the edges of the laminate were additionally fixed.

4 Results and Comparison of Methods

Numerical simulations were carried out in the ABAQUS program. From the qualitative comparison presented in Fig. 3, one can notice the highest similarity of the first model with the real system. The distribution of plastic energy dissipation for the solid and laminate model is similar. In the first model, the stress distribution corresponds to the loads carried by the primary fibers. Dissipation of plastic energy by an order of magnitude smaller in a layered model may result from high dissipation of elastic energy in the form of a large deflection of external layers.

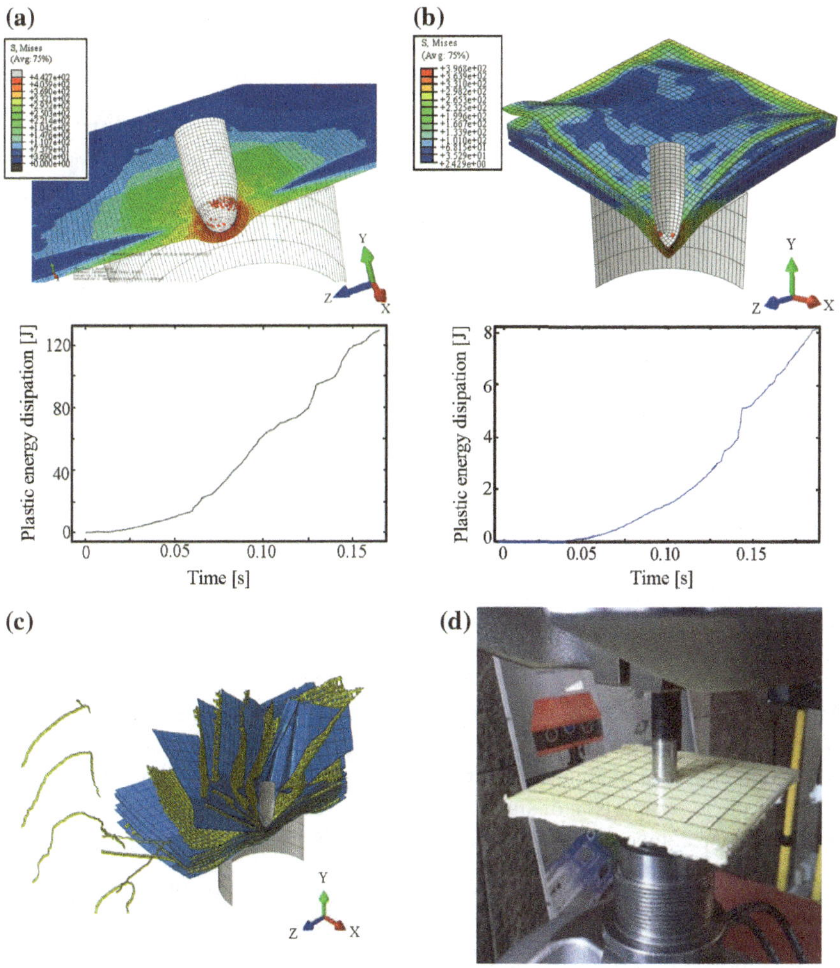

Fig. 3 Qualitative comparison of numerical simulations with the real system for a quasi-static puncturing test: **a** system with a solid laminate model, **b** system with a laminate layered model, **c** system with a laminate layered model that takes into account fabric structure, **d** real system

On the basis of the qualitative analysis of the second and third model, unreal deformations resulting from the lack of edge restraint of the laminate layers were found, while the interlayer adhesion forces were omitted. For further consideration of the impact punching phenomenon, a laminate layered model with edge restraint was used (Fig. 4).

The comparison of the qualitative simulation and the real system at the puncture site shows similar distribution of primary fibers and the post-stroke cone. On this basis, it can be assessed that the following model fully maps the phenomenon of impact breakdown for the tested speed.

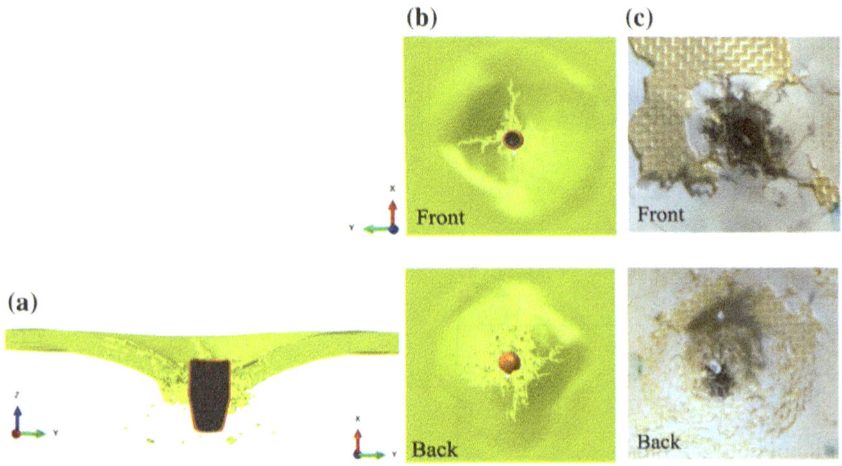

Fig. 4 Comparison of the numerical simulation of the impact break with the real system: **a** cross-section of the layered numerical model of the laminate, **b** front and back view of the layered numerical model of the laminate, **c** front and back view of the real system

5 Conclusions

After conducting numerical analyzes it can be argued that the appropriate adoption of the mathematical-physical model and the mapping of the laminate structure are effective methods of qualitative description of the material in question. On the basis of a qualitative comparison of simulation results to the real system, for puncturing a polypropylene composite layer reinforced with aramid fabric, it is suggested to use:

- the volumetric modeling method with global homogenization of material properties in case of lack of fixing of the laminate model,
- the volumetric modeling method with layered homogenization and lack of adhesion using edge restraint.

Layered or volumetric modeling taking into account the structure of reinforcement requires determining the adhesion between the bodies that make up the composite model or the total restraining of the model's layer edges in order to obtain results consistent with reality. Otherwise, the simulation results do not reflect the actual destruction mechanism. In subsequent works the rotary motion of the projectile will be considered.

Acknowledgements Calculations have been carried out in Wroclaw Centre for Networking and Supercomputing (http://www.wcss.pl), grant No. 452.

References

1. García-Castillo SK, Sánchez-Sáez S, Barbero E (2012) Nondimensional analysis of ballistic impact on thin woven laminate plates. Int J Impact Eng 39:8–15
2. Pach J, Pyka D, Jamroziak K, Mayer P (2017) The experimental and numerical analysis of the ballistic resistance of polymer composites. Compos B Eng 113:24–30
3. Gogineni S, Gao XL, David NV, Zheng JQ (2012) Ballistic impact of Twaron CT709 plain weave fabrics. Mech Adv Mater Struct 19:441–452
4. Kurzawa A, Pyka D, Pach J, Jamroziak K, Bocian M (2017) Numerical modeling of the microstructure of ceramic-metallic materials. Procedia Eng 199:1495–1500
5. Baranowski P, Malachowski J (2015) Numerical study of selected military vehicle chassis subjected to blast loading in terms of tire strength improving. Bull Pol Acad Sci 63(4):867–878
6. Parsons EM, Weerasooriya T, Sarva S, Socrate S (2010) Impact of woven fabric: experiments and mesostructure-based continuum-level simulations. J Mech Phys Solids 58:1995–2021
7. Tan VBC, Ching TW (2006) Computational simulation of fabric armour subjected to ballistic impacts. Int J Impact Eng 32(11):1737–1751
8. Chocron S, Figueroa E, King N (2010) Modeling and validation of full fabric targets under ballistic impact. Compos Sci Technol 70(13):2012–2022
9. Nilakantan G, Keefe M, Bogetti TA (2010) On the finite element analysis of woven fabric impact using multiscale modeling techniques. Int J Solids Struct 47(17):2300–2315
10. Chailleux E, Davies P (2003) Modelling the non-linear viscoelastic and viscoplastic behaviour of aramid fibre yarns. Mech Time-Depend Mater 7(3–4):291–303
11. Touati D, Cederbaum G (1998) Postbuckling of non-linear viscoelastic imperfect. Part I: Material considerations. Compos Struct 4:33–41
12. Wang K (2016) Calibration of the Johnson-Cook failure parameters as the chip separation criterion in the modelling of the orthogonal metal cutting process. Master thesis. McMaster University, Hamilton Ontario
13. Mayer P, Pyka D, Jamroziak K, Pach J, Bocian M, Experimental and numerical studies on ballistic laminates on the polyethylene and polypropylene matrix. J Mech, 1–11. https://doi.org/10.1017/jmech.2017.103
14. https://www.teijinaramid.com/en/product-brochure-twaron/ [online: 2017.10.19]

Ballistic Impacts Experiments and Plastilina/Protection Systems' Modeling

Lionel Gilson, Luc Rabet, Abdellatif Imad and Frederik Coghe

Abstract This work deals with modeling the interactions between projectile, protection and plastilina backing during a non-perforating ballistic impact. For its mechanical behavior, ballistic plastilina was considered as a fluid. A capillary rheometry analysis over a wide range of shear rates was performed to obtain its dynamic viscosity. The mechanical response of ballistic plastilina was deduced from the rheology of plasticine using a strain rate based power law formalism. Based on the spherical cavity expansion theory, it was predicted that ballistic impact is capable of generating very high strain rates behind the impacted region. An experimental study assessed the influence of the impact velocity and the rigidity of the protection on the ballistic response of plastilina blocks, depending on indentation depths of the plastilina behind the impact region. Numerical simulations of the ballistic impact response of the plastilina covered by different ballistic protections allowed the assessment of the dynamic behavior of such systems. Good correlations were obtained between experimental and numerical analysis. It turned out that indentation depths obtained numerically are generally lower than for the experimental measurements.

Keywords Ballistic impact · Plastilina system · Capillary rheometry · Numerical modelling

L. Gilson (✉) · L. Rabet
Departement of Civil and Material Engineering, Royal Military Academy, Avenue de la Renaissance 30, 1000 Brussels, Belgium
e-mail: lionel.gilson@rma.ac.be

A. Imad
Unité de Mécanique de Lille, Université de Lille, Avenue Paul Langevin, 59655 Villeneuve d'Ascq, France

F. Coghe
Department of Weapon Systems and Ballistics, Royal Military Academy, Avenue de la Renaissance 30, 1000 Brussels, Belgium

R. Fangueiro and S. Rana (eds.), *Advanced Materials for Defense*, 4, Springer Proceedings in Materials, https://doi.org/10.1007/978-3-030-34123-7_14

1 Introduction

Most ballistic laboratories use ballistic plastilina as a witness material to test the performance of lightweight protections [6] according to test standards such as NIJ [9]. Indeed, plastilina keeps the mark of a non-perforating impact on a target material. Prior to ballistic testing, plastilina has to be calibrated using a drop test procedure [6, 9]. The calibration and ballistic testing procedure involves measuring the depth of the depression generated by the impacts. Ballistic protection is considered effective when this depth is not greater than a maximum value, generally of the order of 44 mm [5]. Ballistic plastilina belongs to a group of products having a thixotropic behavior [1, 8, 12]. The modeling of the mechanical response of ballistically impacted plastilina remains therefore difficult because of its highly nonlinear and non-Newtonian fluid behavior. A material model for such a behavior can be deduced from the dynamic viscosity obtained from capillary rheometry [3]. Indeed, the rheology of plasticines allows introducing the following relationship in which σ is the stress, η the viscosity and $\dot{\varepsilon}$ the strain rate. For incompressible fluids, strain rate is considered equivalent to the shear rate [3, 10, 12]:

$$\sigma = 2\eta\dot{\varepsilon}$$

However, fitting a complete material model based on such an analysis remains a challenge, especially over the range of strain rates covered by impact processes [2, 6, 7]. Based on the spherical cavity expansion theory, the following relation has been found between the average strain rate $\dot{\varepsilon}_{average}$, the velocity V and the diameter D of the impactor used for the calibration test for the plastilina [1, 6]:

$$\dot{\varepsilon}_{average} = \frac{4V}{D}$$

In the present work, we extent the estimation of the average strain rate to the case of an impactor composed of a projectile interacting with a ballistic protection. Here, we propose a numerical model to simulate the material response of the plastilina (in the current research red Weible® plastilina was used) from a capillary rheometry analysis. Ballistic tests with three different protection configurations were performed to measure the indentation depth into plastilina. Numerical simulations of these impact processes were elaborated to evaluate the impact response of the system projectile/protection/plastilina [4].

2 Experimental Procedure

The viscosity of plastilina was measured with a RHENOGRAPH capillary rheometer according to the standard ISO 11443:2005 where a controlled temperature of 28 °C had been imposed. The dynamic viscosity and a fitted power law are shown on Fig. 1a.

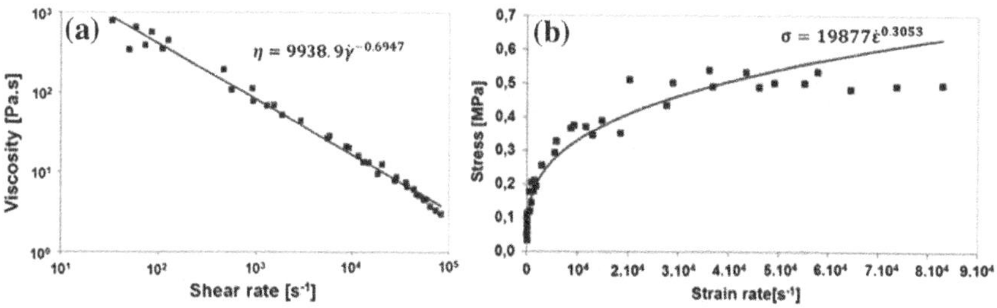

Fig. 1 Experimental results and fitting of, **a** dynamic viscosity as a function of shear rate and, **b** corresponding stress-strain rate relationship

It allows to deduce the stress-strain rate relationship and the corresponding adjusted power law to be used for numerical simulations (Fig. 1b). It is interesting to note (see Fig. 1b) that the adjusted power law fits well experimental values up to about 40,000 s^{-1}. Beyond this value, the response of the adjusted model overestimates the experimental values.

Three configurations of ballistic protections were considered with ballistic plastilina according to the NIJ procedure [9]. The first and most rigid one (E = 68.9 GPa) consisted of three aluminum plates, each having a thickness of 3 mm. The second, orthotropic, with intermediate rigidity ($E_a = E_b = 56.3$ GPa), consists of a 5 mm thick DYNEEMA® HB80 plate (laminate with 34 sub-layers). Finally, the softest orthotropic tested configuration ($E_a = 29.1$ GPa; $E_b = 31.6$ GPa) consisted of ten stacked plies of Kevlar® XP™ S102 with a total thickness of 4.6 mm. The impacts were made using FN commercial projectiles 9 mm diameter with a lead core and a 70/30 brass jacket having velocities ranging from 320 to 430 m/s. The velocities of the projectiles were measured with a 10 GHz Weible® Doppler radar (0.2% accuracy) and a dual optical basis. Figure 2 describes the complete measurement system.

Plastilina being opaque, only the final surface state of the plastilina could be considered in this analysis. Figure 3 presents the typical aspect of the surface of the plastilina after the impact on the Kevlar® based protection at 346.1 m/s.

Taking into account the total thickness of each configuration of protection and the lowest velocity of the projectile (320 m/s), we estimated the average strain rate attainable in the plastilina at the impact region using the spherical cavity expansion theory. We admit that the diameter of the impactor, consisting of the projectile/protection couple, is the sum of the diameter of the projectile and twice the thickness of the protection. This means for the three respective protections that the average strain rates could reach values up to 47,000, 67,000 and 70,000 s^{-1}.

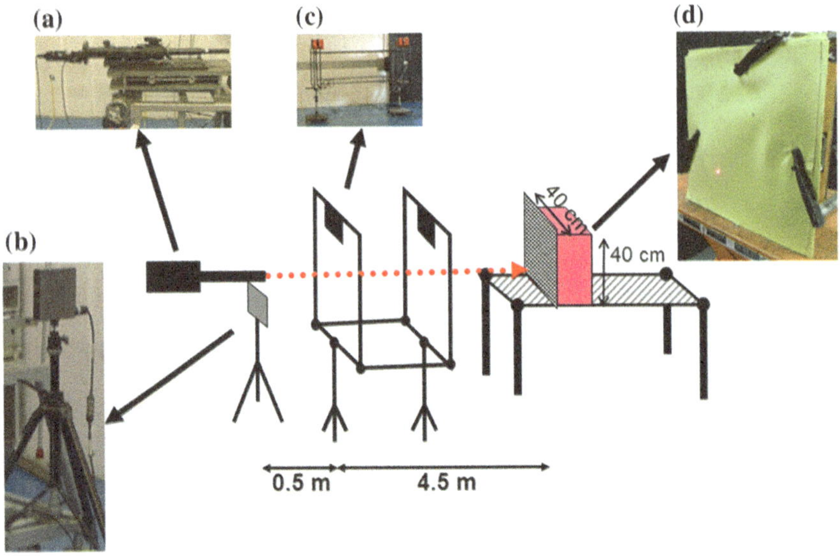

Fig. 2 Experimental setup, **a** launcher, **b** Weible® Doppler radar, **c** optical basis and, **d** fixed Kevlar® based sample on plastilina bloc

Fig. 3 Surface aspect of the plastilina after impact of the Kevlar® based protection

3 Numerical Analysis

Based on the experimental setup, several FEM numerical simulations were performed to evaluate the depressions depth in the plastilina behind the armor after impacts. Figure 4 details the model proposed in the case of the aluminum based protection.

Using symmetry, only a quarter of the models had to be simulated. The mesh of the protection was simply generated from the mesh surface of the modeled plastilina.

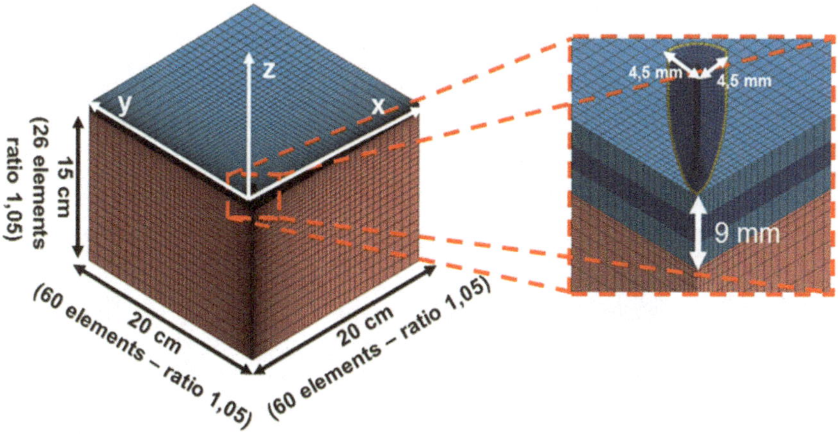

Fig. 4 Description of the numerical simulation in the case of an aluminum based-protection

A progressive mesh was proposed to reduce the number of elements in the simulation while strongly detailing the impact region. The only difference between the models proposed for the representation of the protections is the use of four nodes shell elements to model the layers of the Kevlar® and Dyneema® protections instead of hexahedral (aluminum) solid elements. Great care was taken in the selection of material models for the projectile and protections [4]. The plastilina was modeled according to the results deduced from the capillary rheometry measurements (Fig. 1). Figure 5 presents numerical results for the three types of protection studied in this investigation.

4 Results and Discussions

The main evaluated parameter in this work is the indentation depth of the plastilina behind the impacted protections (Fig. 3). Figure 6 presents the comparison between tests and simulations for all protections configurations and impact velocities.

Figure 6 indicates a good agreement between experiments and simulations. Overall, there is an increase in indentation depth with projectile velocity and with the flexibility of the protection configuration. However, in all cases, the numerical simulations provide a slightly lower penetration depth than that of the tests. This could be explained by the use of the stress-strain rate power law fitted with two parameters to model the plastilina, which overestimate the stresses for large values of the strain rate (about 40,000 s^{-1}). As estimated, when ballistic impacts occur, the average strain rate becomes, at least temporarily, very high in the plastilina. Greater values than 40,000 s^{-1} seem to be easily reached. However, for simulations, the strain rate based power law used for modeling the plastilina has a more rigid behavior when strain rates become high. This could explain why the simulations give values of indentation

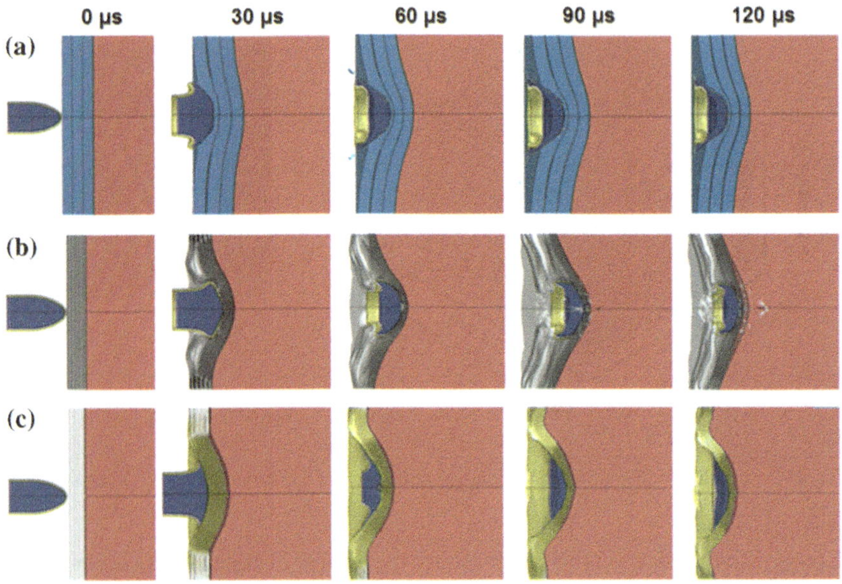

Fig. 5 Deformation evolution during impact process at 350 m/s, with a protection, **a** aluminum, **b** DYNEEMA® HB80 and, **c** Kevlar® XP™ S102

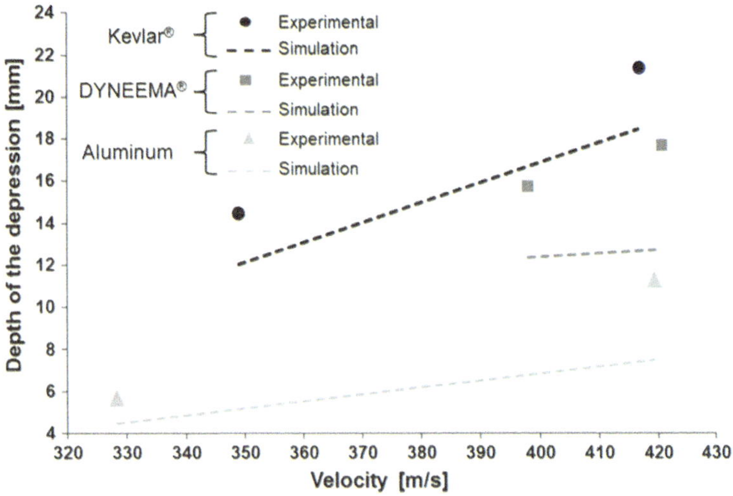

Fig. 6 Evolution of deformation during impact process at 350 m/s, with the three different protections

depth lower than those observed during experiments. Therefore, great care should be taken for modeling complex fluids such as ballistic plastilina when subjected to potentially great strain rates, such as during impact processes. One solution might be to introduce more complex material models for simulations of these materials [11].

5 Conclusions

The mechanical behavior of ballistic plastilina was evaluated using a capillary rheometry analysis over a very high range of strain rates. A material model was deduced to simulate the dynamic response of this thixotropic material. A strain rate based power law with two parameters was selected to fit the response of the plastilina. This choice was made according to the rheology of plasticine. However, such a material model deviates slightly towards higher values of stresses at very high deformation rates. As a result, the adjusted power law becomes more rigid than that measured experimentally by rheometry.

An experimental setup was developed to perform ballistic tests with plastilina together with different configurations of protective materials. The selection of protective materials used in this analysis was based on their respective rigidities. The main parameters measured during the tests were the projectile's velocity and the indentation depth of the plastilina after the ballistic impacts.

As expected, indentation depths increase with projectile's velocity and flexibility of the protective structure. In addition, based on the spherical cavity expansion theory, the average strain rate of the plastilina backing during a ballistic impact was estimated. It was shown that the strain rate can reach very high values, at least for a short period of time. In these conditions, the fitted strain rate based power law with two parameters overestimates the materials' rigidity.

Based on these results, numerical simulations were carried out to model and assess the ballistic impact responses of the different considered systems projectile/protection/plastilina. The different simulations corroborated the experiments. The simulated indentation depths into the plastilina increase with the impact velocity of the projectiles and the flexibility of the simulated ballistic protection. However, all simulations underestimate the indentation depth. The choice of the strain rate based power law with two parameters used for modeling the plastilina is responsible for this effect. Indeed, over a wide range of deformation rates, such a model cannot fit perfectly the real material response of plastilina. It tends to become slightly stiffer than expected according to the capillary rheometry analysis. Therefore, a more in-depth analysis should be considered to model correctly the complex dynamic behavior of thixotropic fluids subjected to high strain rates as is the case for ballistic impacts. More complex material models must be introduced to simulate the ballistic impact response of the plastilina.

References

1. Buchely MF, Maranon A, Silberschmidt VV (2016) Material model for modeling clay at high strain rates. Int J Impact Eng 90:1–11
2. Carton E, Roebroeks G, Broos H, Halls V, Zheng J (2014) Characterization of dynamic properties of ballistic clay. In: Proceedings of personal armour systems symposium (PASS2014), Cambridge. UK
3. Dupuis JF, Villemaire JP, Felder E (1994) Capillary rheometer measurement of a plasticine used in order to simulate the hot extrusion of metals. J Mater Process Technol 43:237–257
4. Gilson L (2017) Etude du comportement mécanique des multi-matériaux soumis à un impact balistique: approches expérimentale et numérique. PhD dissertation. Université de Lille et Ecole Royale Militaire
5. Hanlon E, Gillich P (2012) Origin of the 44-mm behind-armor blunt trauma standard. Mil Med 177:333–339
6. Hernandez C, Buchely MF, Maranon A (2015) Dynamic characterization of Roma Plastilina No. 1 from Drop Test and inverse analysis. Int J Mech Sci 100:158–168
7. Mates SP, Riley M, Forster A, Rice K (2012) Modelling unconfined dynamic compression tests of Roma plasticine #1. In: Proceedings of personal armour systems symposium (PASS2012), Nurenberg, Germany
8. McClay KR (1976) The rheology of plasticine. Tectonophysics 33:T7–T15
9. NIJ Standard-0101.06 (2008) Ballistic resistance of body armor. National Institute of Justice, US Department of Justice
10. Schöpfer MPJ, Zulauf G (2002) Strain-dependent rheology and the memory of plasticine. Tectonophysics 354:85–99
11. Vingaard B, Endelt B, deClaville J (2007) Implementation of a material model with shear rate and temperature viscosity. In: Proceedings of the 6th European LS-DYNA users conference
12. Zulauf J, Zulauf G (2004) Rheology of plasticine used as rock analogue: the impact of temperature, composition and strain. J Struct Geol 26:725–737

Thermal Isolation/Barriers for Small Caliber Weapon Applications

Adam L. Foltz, Adam Jacob and Laurie Florio

Abstract The research conducted under this effort aimed at analyzing thermal barrier concepts intended to reduce visual signature of small caliber weapon systems (barrels, suppressors, etc.) resulting from high rates of firing. Analysis of heat transfer mechanisms and modeling and simulation of thermal barrier concepts were conducted. Basic principles were explored to identify limitations of application conditions and influencing system design parameters. Modeled results representing live fire test cases were in good correlation with both standard M4A1 test data and vacuum barrier results. Barrier effectiveness was shown to decrease with higher heating rates as barrel temperatures increase and radiation exchange between barrel and barrier rises.

Keywords Thermal barrier · Small caliber weapons · Modelling · Computational fluid dynamics

1 Introduction

Aggressive firing rates can result in extreme high temperatures in barrels and suppressors. To date there has been minimal research in the area of technologies intended to dissipate heat or mitigate build-up during cook-off/aggressive firing scenarios.

Historically our enemies have had an inability to fight at night and have relied on weapon muzzle flash to provide an ad hoc aiming reference during ambushes and traditional fire fight situations. Advancements in thermal imaging technology continue to lower system costs giving our enemies improved detection capabilities at night, which can jeopardize mission effectiveness and soldier survivability. Smart phone applications and UAV mounted systems are becoming more prevalent, identifying the critical need for developing thermal barrier concepts for small caliber weapon applications (Fig. 1).

A. L. Foltz (✉) · A. Jacob · L. Florio
US Army ARDEC, Picatinny Arsenal, NJ 07806, USA
e-mail: adam.l.foltz.civ@mail.mil

R. Fangueiro and S. Rana (eds.), *Advanced Materials for Defense*, 4, Springer Proceedings in Materials, https://doi.org/10.1007/978-3-030-34123-7_15

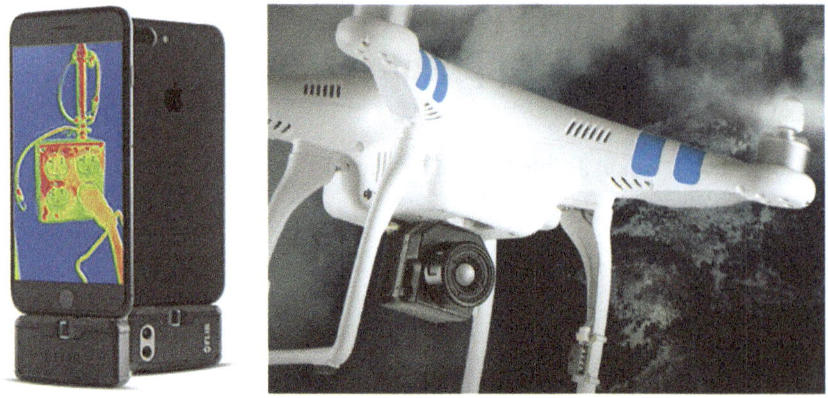

Fig. 1 FLIR products [1]

The research area for this effort primarily focused on vacuum barrier concepts, barriers which both reflect and diffuse heat, reduce heat transfer across the barrier by absence of atmosphere, and only allow end area conduction.

Particularly of interest was Insulon®, a technology developed by Concept Group [2]. An Insulon® Shaped-Vacuum™ Thermal Barrier uses a vacuum layer to form a virtually impenetrable barrier to the conductive transfer of thermal energy. It has flexibility in size and shape to accommodate unique thermal design challenges. It blocks the conduction of heat or cold or works in reverse to protecting sensitive equipment, components, fluids or materials that need to operate at optimal temperatures even though they are surrounded by extremely high or low exterior temperatures, from −321 to 2000 °F. It requires 40× more fiberglass to have similar conduction to a 0.1 mm Insulon thermal barrier, 20× more expanded polystyrene, and a 25× larger air gap [2]. Applications include batteries and fuel cells, automotive, military and defense, aerospace (component isolation), consumer goods (e-cigarettes), and medical devices (transport, catheters, spectroscopy, cryogenics) (Fig. 2).

In order to investigate the basic principles behind the thermal barrier concept, the conditions under which it may hold more promise, and the range of possible improvement in the thermal state of a system containing a barrier over a system without the barrier, a number of sets of computational simulations were conducted. All simulations were conducted using ANSYS Fluent Computational Fluid Dynamics (CFD) software.

2 Modeling and Simulation

In order to investigate the basic principles behind the thermal barrier concept, the conditions under which it may hold more promise, and the range of possible improvement in the thermal state of a system containing a barrier over a system without the

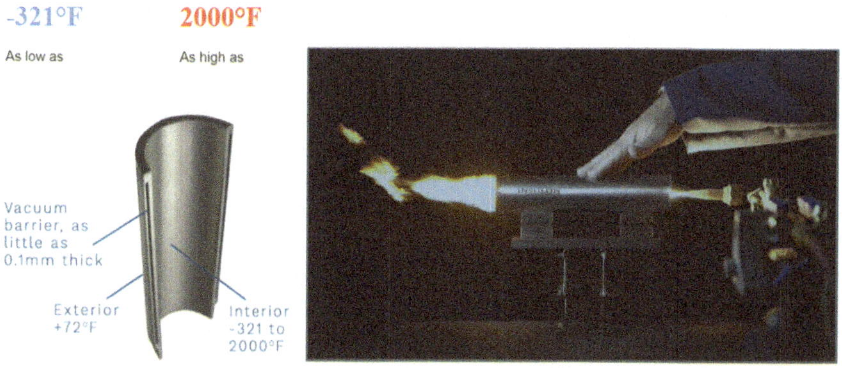

Fig. 2 Insulon performance [2]

barrier, a number of sets of computational simulations were conducted. The simulations vary in complexity starting with a simple heat flux applied to a material with just an air gap and progressing to a multi-shot firing schedule applied to gun tube with a thermal barrier shield. All simulations were conducted using ANSYS Fluent Computational Fluid Dynamics (CFD) software [3] and methods and results from previous modeling efforts [4, 5]. Specific studies are described below:

Study 1—Constant heat flux, solid tube versus air gap

The objective of this preliminary study was to compare the thermal conditions that develop when a heat flux is applied to the inner surface of a cylindrical tube for the cases of a solid tube and a tube that contains an air gap of uniform thickness. In this way, the potential effects of a thermal barrier and the basic phenomena at work in the barrier system could be identified (Fig. 3).

The results of this study illustrate that while the air gap delays the increase in the temperature at the outer surface of the tube, since the radially directed heat flow is slowed by the more resistive air gap within the barrier, the temperatures at the inside surface of the barrier are generally higher than if no barrier is present. This simple model also illustrates that the main mechanism through which this thermal barrier functions is to restrict the major paths of the flow of heat to the conductive heat flow through the small solid zones at the left and right of the barrier and radiative exchange across the barrier surfaces with minimal conduction through the air gap.

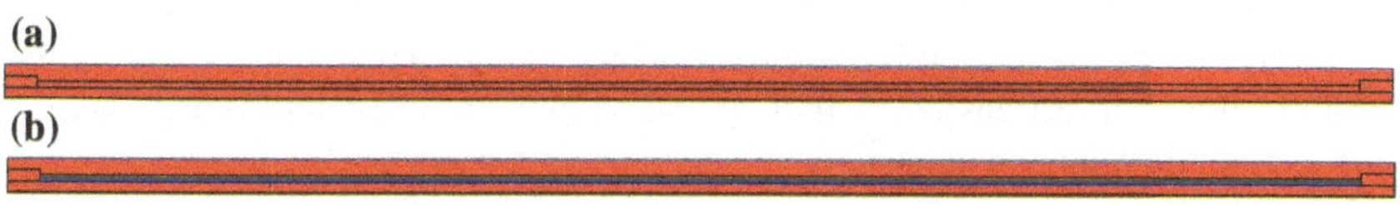

Fig. 3 Model setup **a** solid tube, **b** air-gap within the tube thickness

Study 2—Constant heat flux, emissivity influence

Next, the emissivity of the inner barrier gap surfaces was varied. The one large barrier from Study 1 and a case with multiple smaller barriers was investigated. The results of the Study 2 cases continue to indicate the "thermal barrier" created by the air gap restricts the rate of heat flow in the radial direction, delaying the heating of the outer portions of the tube. The air gap provides greater resistance to heat flow across the thickness of the heated tube compared to a completely solid tube. Further, cases with lower emissivity and a long, uninterrupted gap space produce a better barrier to the heat flow than higher emissivity barrier surfaces and barriers with multiple conductive heat flow paths between multiple gaps. Ultimately the results demonstrate a less effective thermal barrier with lower reflectivity/higher emissivity values.

Study 3—Constant heat flux, air gap versus vacuum

In the third study, the thermal conditions with a vacuum in the barrier volume filled with air in the previous studies was considered. The same geometric configuration with the solid tube and the tube with a single barrier gap were studied. The emissivity of the barrier surfaces was also varied including the potential for two different emissivity's, one on the top surface of the gap and one on the bottom surface of the gap. Also included was a case with no radiation in the gap, taken as an upper limit on the potential barrier performance.

Based on the results of the investigation, a number of important temperature field characteristics can be identified. As air is replaced by a near vacuum, the barrier effectiveness increases, with reduced temperatures in the solid material outside the barrier. However, the drop in exterior temperature comes at the expense of higher temperatures in the interior of the tube. The sensitivity of the thermal conditions to the emissivity with the vacuum barrier follows trends similar to those for the air filled gap. The differences between the conditions that result with an air and vacuum filled gap decreases as the surface emissivity increases and the radiation exchange across the gap grows. For the inner side of the barrier, higher surface temperatures are seen for the vacuum barrier case vs. the air filled gap due to the resisting heat flow. For this case, the boundary condition of no radiation represents the upper bound of the barrier performance. For the outer barrier surface, the rate of increase in temperature for the vacuum barrier eventually surpasses the solid tube case due to radiation.

Study 4—Conditions for varied outer surface tube boundary conditions

The next step in the investigation varied the boundary conditions at the outer surface of the tube to examine how these boundary conditions influence the effectiveness of the barrier. For one set of studies, an insulated boundary condition set was employed at the outer surface of the tube so that no heat was removed from the outside surface. In the other set of studies, a mixed convection and radiation boundary condition was applied where a heat transfer coefficient and an ambient temperature were applied with a common radiative emissivity.

For the insulated outer boundary condition, no heat flow occurs, resulting in temperatures that continue to rise, demonstrating negligible barrier benefits with long duration heat loading. With the mixed convective and radiative outer surface boundary condition, a steady temperature distribution was reached within the tube. The time to reach the steady state was longer for the cases with the lower emissivity where the rate of heat transfer is lower.

Overall results showed that for a high input heat rate and long duration input time period with lower levels of heat removal from the outer surface of the barrier, the benefits of the barrier are questionable. For lower levels of heat input and short durations, the higher temperatures at the inner vacuum portions may be less problematic. Removal of the energy "trapped" in the interior portions of the tube may reduce the potential issues for the higher inner tube portion temperatures and might also reduce the heat transfer to the outer portions of the tube.

Study 5—Constant heat flux, forced cooling fluid between barrier/solid

The potential issues related to the hotter inner tube temperatures resulting from the implementation of the barrier technique prompted an investigation into the feasibility supplementing any cooling by forced convection cooling for a system consisting of a solid tube, a barrier shroud, and a concentric forced convection flow path between the solid tube and the barrier. The objective of this fifth Study was to determine the sensitivity of the following design parameters on the thermal conditions that result when air or water flow is used to cool a heated tube with a concentric thermal vacuum barrier: the properties of the working fluid: air or water, velocity of the cooling fluid, the distance of the inner surface of the concentric thermal vacuum barrier, and the thickness of the vacuum in the thermal vacuum barrier. The thermal loading on the heated tube, the heated tube dimensions, and thickness of the solid material forming the barrier gap were held fixed. The system to be investigated and some important parameters are depicted in Fig. 4.

A total of sixty-three different configurations were simulated for the same thermal loading at the inner surface of the tube and the same tube geometry. For the air, three different air cooling velocities were studied at 20, 40, and 60 m/s. For the water, velocities that produce the equivalent mass flow rate as the air were applied at 0.02344, 0.046884, and 0.070327 m/s. For each of these velocities and fluids, three different spacing between the outer heated tube surface and the inner surface of the barrier were investigated at 3, 6, and 9 mm. Then, for each fluid, velocity, and tube-barrier spacing cases, three vacuum gap spaces in the thermal vacuum barrier of 0.508, 1.016, and 2.032 mm were applied to the system. Additionally, a case with a heated tube with no barrier and cases for each geometric configuration with ambient zero velocity air were simulated to form a baseline for comparison.

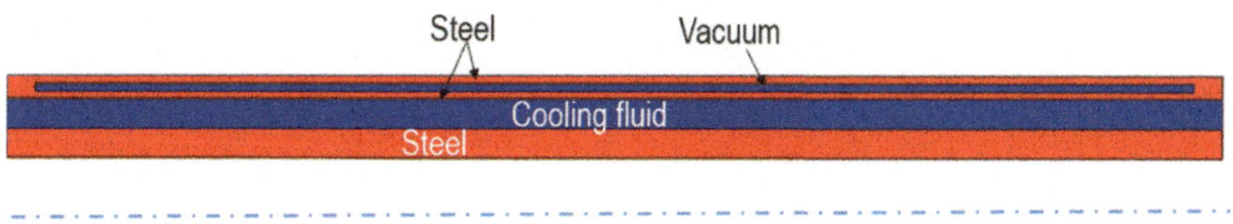

Fig. 4 Model setup study #5

With water as the cooling fluid, the conditions found differ significantly due to the greater heat capacity of the water and the different radiative properties. The temperatures at the outer surface tend to decrease with higher flow velocities, but the decrease appears to diminish with increases in the water velocity. The temperatures eventually cease increasing indicating the heat input to the tube is being removed by the water flow. The temperatures reached at the outer surface of the tube are significantly lower compared to the air cooled cases.

In each case with air or water, increasing the spacing of the cooling fluid flow path results in a slight increase in the temperature at later times. Perhaps the higher initial or net heat flow rate into the fluid results in a slight decrease in the heat outflow as the fluid is warmed and therefore the higher temperatures at the outer tube surface (Fig. 5).

Study 6—Complex heat load to replicate live fire

In the last investigation set, a heat load more representative of live fire conditions for a weapon system were applied. A 210 round firing sequence at 120 rounds per minute was applied to the M4 system. The barrel and barrier temperatures (average surface temperatures) are traced over time during the time period when firing is taking place as well as after firing has stopped and the gun system is cooling. The simulations were run until the exterior or outer barrier surface reaches near its maximum temperature. Live fire testing was conducted using a baseline weapon configuration and a suppressed weapon configuration, both with and without a vacuum barrier to compare with the modeled results. Peak temperature results are shown in Figs. 6 and 7.

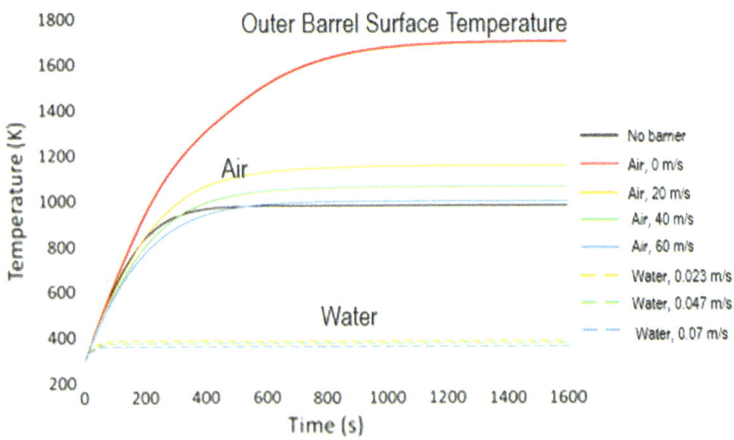

Fig. 5 Out barrel surface temperature (study #6)

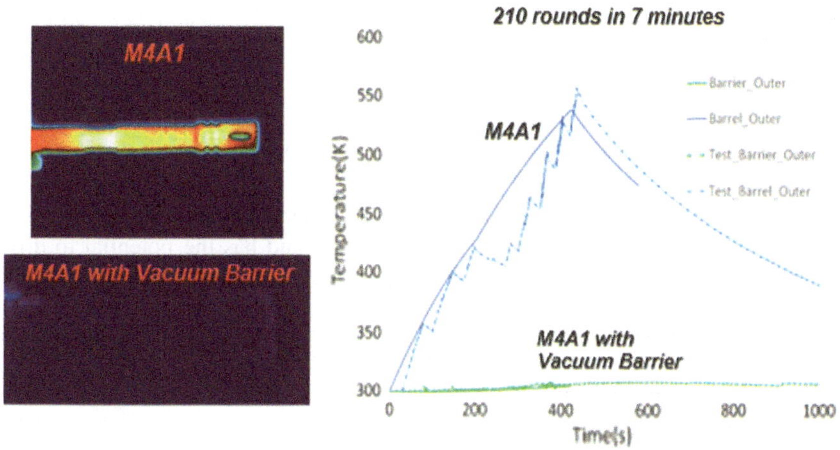

Fig. 6 Test results M4A1

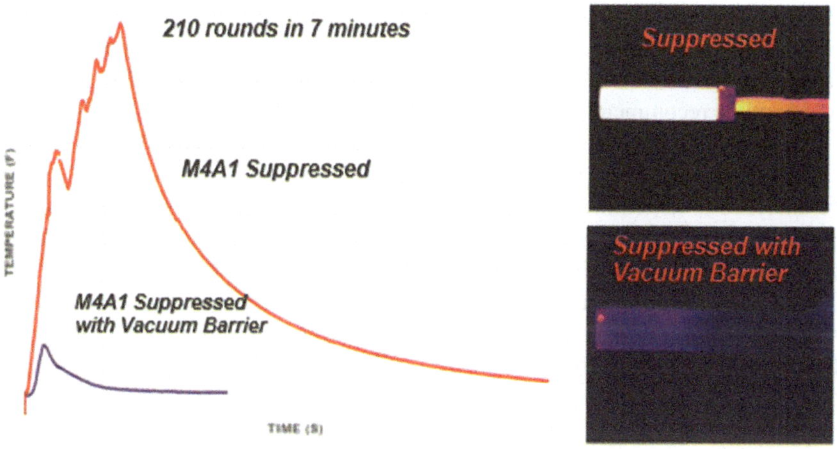

Fig. 7 Test results M4A1 suppressed

3 Results and Conclusions

Simulations varied in complexity starting with a simple heat flux applied to a material with just an air gap and progressing to a multi-shot firing schedule applied to gun tube with a thermal barrier shield. Studies included: (1) Solid tube versus air gap, (2) Emissivity influences, (3) Air gap versus vacuum barrier, (4) Varied boundary conditions, (5) Addition of forced air/water, and (6) Complex heat loading to replicate live fire testing. The complex heat load models were setup to directly compare results to live fire test data. An approximate location to the measurement position was selected for data comparison and the results are shown in Figs. 6, 7. A good correlation between the standard barrel test data and the model results for a mixed convection boundary condition case can be observed. Also, the good correspondence between the barrier covered barrel results and the model results for the stagnant air case can be observed. The modeling results presented here provide a reasonable estimation of the thermal conditions that develop with the barrier and with the baseline standard barrel system.

The results of this effort generated a greater understanding of thermal loading and heat mitigation in gun barrels and suppressors and has the potential to drive science and technology based material solutions for gun barrel/suppressor thermal management and increased maintainability.

4 Recommendations

Although significant thermal signature was achieved for the firing scenarios tested, a more aggressive firing cadence with high heating loads could result in a less effective thermal barrier and may have other negative system performance impacts (i.e. barrel wear, decreased accuracy, increased dispersion) due to the increased temperatures on the inside of the vacuum barrier. Additional testing and analysis would be necessary to determine recommended operating applications and limitations.

Further modeling should be conducted on varying heating loads/firing conditions to help define potential optimal operating ranges. Outer surface temperatures should be analyzed to determine thermal signature performance and bore surface temperatures should be monitored. The effects of geometric parameters, placement and overall barrier configuration should be studied to determine effects on performance. Also the strength and harmonics associated with the barrier configuration and gap spacing should be explored. Surface conditions and materials for the outer barrel surface and the inner bore surface should be studied, including those used for support connections along the tube length.

Although modeled, a basic air gap barrier was not tested. Testing should be conducted to compare an air gap barrier of similar geometry vs. the vacuum barrier prototype to further characterize the significance of the vacuum barrier concept.

Finally additional means for removing heat, particularly from the inside volume of the barrier should be investigated further. The energy input must still be removed though it may be distributed differently, either through additional conductive pathways, possible phase change materials, or convective cooling enhancements.

Acknowledgements This work was funded by the ODOT under the FY17 Systems Concept and Technology (SC&T) effort. The authors would like to acknowledge the following people for providing information and data used in the development of this work: Howard Kent, David Reid, Mike, Scott, Louis Schnibbe, Tuhin Chatterjee, Dan Cler, Andy Littlefield, Mikhail Kats, Thomas Miskovich, Anthony Cautero, and Johnathan Parker.

References

1. FLIR Systems. Retrieved from http://www.flir.com/home/
2. Insulon Technology. Retried from https://conceptgroupinc.com/applications/
3. ANSYS Fluent
4. Florio LA (2008) Modeling heat transfer due to the firing of one round in an M4. ARWSE-TR-08005
5. Florio LA (2009) Thermal conditions in M4 barrel resulting from repeated firing. ARSWE-TR-09012

Development of a New Technology to Manufacture an Additional Ballistic Protection Panels (Add on) for an Armored Personnel Carrier

Julio Sérgio Dolce da Silva, Juliana Machado da Mota, Liliana Burakowski, José M. Matias Lopes and Renan Munhoz Rodrigues

Abstract This work deals with the development of a STANAG 4569 level III PLUS (12.7 × 99 mm AP M2) ballistic protection of ceramic panels (add on) for a Brazilian armored personnel carrier. The research started with the study of the physical characteristics of different hexagonal ceramic chips and their best assembly on the plate. After the selection of the best ceramic chip the final configuration of the ballistic panel was completed following the procedures according to NATO STANAG AEP-55—volume 1 [1]. The ballistic test panels were produced by "ALLTEC Indústria de Componentes em Materiais Compostos", in Brazil and the ballistic tests were carried out by the "Centro de Avaliação do Exército (CAEx)", the Brazilian Army Proving Ground, in Rio de Janeiro. After many ballistic tests led by the CAEx technical staff, finally an assembled panel was produced by ALLTEC with the choice of ceramic chip configuration, as well as a special bonding material and the backing support in aramid fiber fabric. The best results achieved led the researchers to produce a prototype panel that was approved after exhaustive tests carried out by CAEx to consolidate the results and to approve the final product. The homologation and validation of the entire additional ballistic protection panel (ADD ON) was homologated by the Brazilian Army to be applied on its personnel armored carrier.

Keywords Ballistic protection · Ceramic panels · 12.7 mm AP ammunition · Personal carrier vehicle

J. S. D. da Silva (✉) · J. M. Matias Lopes
Brazilian Army, Rio de Janeiro, Brazil
e-mail: jsdolce@gmail.com

J. M. Matias Lopes
e-mail: mljosem@gmail.com

J. M. da Mota · L. Burakowski · R. M. Rodrigues
Alltec Indústria de Componentes em Materiais Compostos Importação e Exportação Ltda, São Paulo, Brazil
e-mail: juliana.mota@allteccomposites.com.br

L. Burakowski
e-mail: liliana@allteccomposites.com.br

R. M. Rodrigues
e-mail: renan.munhoz@allteccomposites.com.br

R. Fangueiro and S. Rana (eds.), *Advanced Materials for Defense*, 4, Springer Proceedings in Materials, https://doi.org/10.1007/978-3-030-34123-7_16

1 Introduction

The present work deals with the development and manufacture of an additional shielding (add on) to the structural ballistic steel that composes the hull of an armored vehicle on wheels of personnel carrier, so that it is able to withstand shots of perforating projectiles 12.7 mm AP M2.

The additional ballistic armor that has been developed is composed of ballistic ceramics adhered to a backing (composite based on aramid fibers). The internal parts of the hull are covered with spall liner panels made with fabric aramid to protect the crew against fragments that can detach from the external projectiles impact.

In order to the shielding system acts properly, the projectile tip must be partially destroyed and thus it is indicated that part of the shield should be composed of high hardness material, such as ceramic.

The shielding panels composed of ceramic elements presents an enormous potential of use due to its high hardness able to reduce the velocity of the projectile and turns it into small fragments that are absorbed by the flexible layer, composed by polymeric material and fibers, that supports it.

2 Materials and Methods

In the present work, ceramic elements, adhesives and sealants were used to make the panels of the additional shield and composite material was used to make the final assembly of the whole package, forming a lightweight and compact hybrid ballistic panel.

An extensive study was carried out, aiming to find the best combination of materials to compose the internal and external structure of the shielding panel in order to achieve a great performance in protection against 12.7 mm AP M2 projectiles.

The ballistic panels were tested in the ballistic tunnel of the Brazilian Army Evaluation Center (CAEx), following NATO STANAG AEP-55 and the results showed that the materials chosen to manufacture the ballistic panel are the ones that presented the best performance.

Up to the final configuration of the shielding panel kit, as it was approved by CAEx, an extensive study included several test steps for choosing the materials that assembled the whole pack: ceramic, backing and envelope.

2.1 Ceramics

To manufacture the shielding panel, alumina was chosen as the high hardness element. Among the ceramics, alumina is one of the less cost-benefit value materials. Alumina-based ceramic chips were purchased from *Coorstek do Brasil Materiais*

Table 1 Properties of alumina

Property	
Alumina content	99.5%
Density	3.82–3.86 g/cm^3
Modulus of elasticity	360 GPa
Vickers hardness	13–15 GPa
Flexural stress	370 MPa
Compressive stress	2600 MPa
Tenacity fracture	4–5 MPa m$^{1/2}$

Avançados Ltda. and have, according to the supplier, the properties are shown in Table 1.

2.2 Adhesive

The type and state of an adhesive, its properties, the temperature and the time required for its cure influence the selection of the best adhesive for the production of a structure more suitable to the fixation of the other elements of the shielded panel. These factors and others, belonging to the material itself and the shape of the structure influence the method of application and fixation of the adhesive, as well as its assembly and installation. For this task, many adhesive joints composed of rigid and flexible substrates actual used in the shielding panels, were analyzed in accordance with ASTM D6862-04, after carried out tests on several adhesives, the best result was obtained with the epoxy resin (Fig. 1).

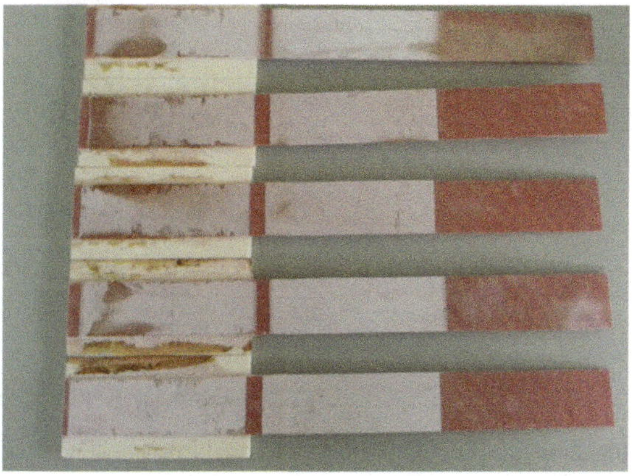

Fig. 1 Adhesive seals used in the shielding panels

2.3 Envelope

The technique used to envelop the shielding panel consists of sealing the entire structure and all the edges, without leaving any cracks for moisture infiltration or any other products to penetrate inside.

The objective of the accelerated aging test was to verify the behavior of the adhesive interface to know if there were changes in the composite material according to the climatic conditions, simulating the different situations to which the armored vehicles are subjected in real operations.

In this context, both the adhesive joints and the composite mini-panels (Fig. 2) were placed in a simulating chamber of accelerated aging with the application of moisture, temperature and UV-A simultaneously, in order to make the evaluation of the influence of weathering on the adhesive joints and envelope. The test was conducted for a period compatible with 120 months of field exposure, and no structural changes were observed either in the joints or in the envelope.

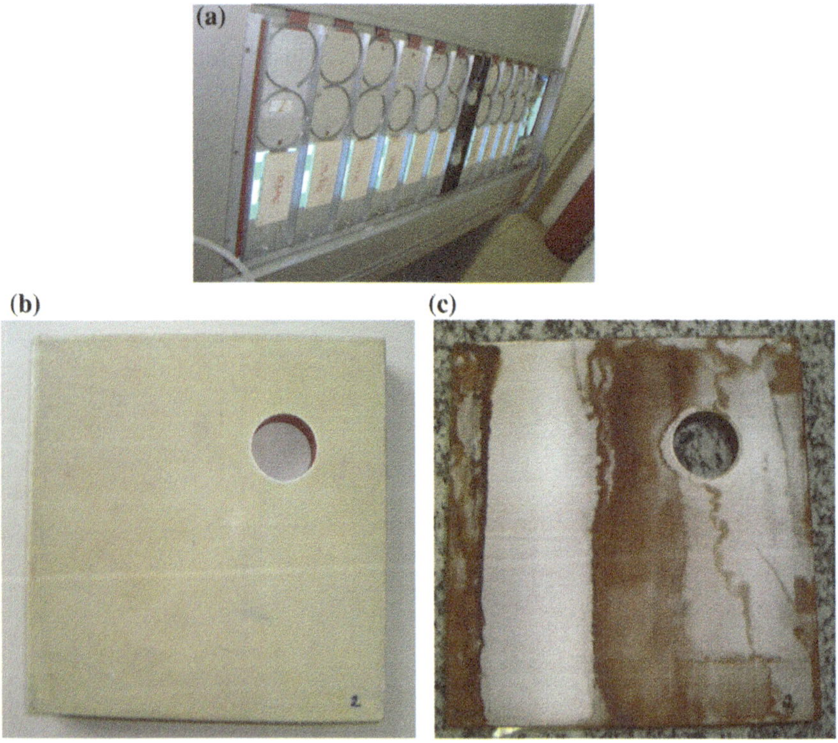

Fig. 2 **a** Test of adhesive joints and **b** and **c** panels aging

2.4 *Manufacture of Test Specimens for Ballistic Tests*

The fabrication of the specimens involved three distinct stages:

Step 1: Manufacture of a template with geometry compatible with the dimensions of the specimens, insertion of the ceramic elements in the template and addition of the adhesive.
Step 2: Manufacturing the backing, consisting of layers of aramid fabrics, polymer film and adhesive laminates. The set was autoclaved for curing, under controlled pressure and temperature.
Step 3: After the previous steps the finishing process was followed, with the integration between the ceramic elements and the backing. This process consists of making the envelope itself and of wrapping the panel. In a template the fiber fabrics were laminated together with a polymeric film. Then the ceramic + backing package was placed into the envelope template and closed, then the whole assembly was autoclaved again.

2.5 *Panel Validation Tests*

In order to verify the behavior and integrity of the panels, they were subjected to shots in a multi-hit condition (double shooting).

Ballistic tests were performed in the CAEx ballistic tunnel and no complete penetration of the projectile was identified in the structural ballistic steel samples positioned behind the panel, that represent the steel wall of the armored car.

Figure 3 shows two of the panels subjected to 12.7 mm AP M2 projectile firing in multi-hit condition. One can observe the proximity of double shooting and the integrity maintained in the vicinity of the perforations.

Representative panels were manufactured for each area of the vehicle that required additional protection against 12.7 mm AP M2 ammunition. Firstly the drawings of each piece were made in CAD including all the details of drilling and composition.

Figure 4 shows two panels tested in the ballistic tunnel, with all the items necessaries for the assembly them on the vehicle, including with the "dog-bones", that provide protection in the junction areas of the panels.

3 Conclusion

The present work presented the different stages to design and manufacture the additional external ballistic armor kit for a medium armored personal carrier vehicle on wheels, as well as the entire process of research, material selection and supplier chain development, in order to manufacture ceramic ballistic plates, besides its aging tests, the technical evaluation tests and military operational field approval tests.

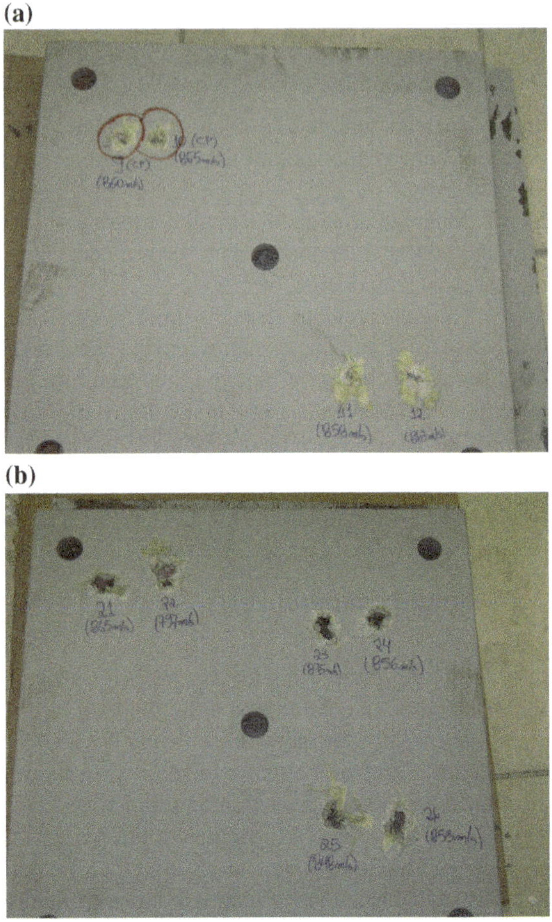

Fig. 3 Multi-hit ballistic tests

For the success of this project, partnerships with the Brazilian Army's science and technology military organizations, such as the Directory of Manufacturing (DF) and the Brazilian Army Proving Ground Center (CAEx), were established, as well as partnership with an institution for the promotion of research as the Financier of Studies and Projects (FINEP), which fully funded this project and contributed to obtaining a quality final product that achieved full approval in the ballistic and operational tests.

All this process was also used for the technical training of Alltec's engineers, researchers and staff, for the development of materials for the defense area, with a special focus on ballistic armor and also for the development of new suppliers for this production chain.

The planning, design, development and execution of the manufacturing process and the final homologation of the complete ballistic protection kit encompassed four stages.

(a)
(b)

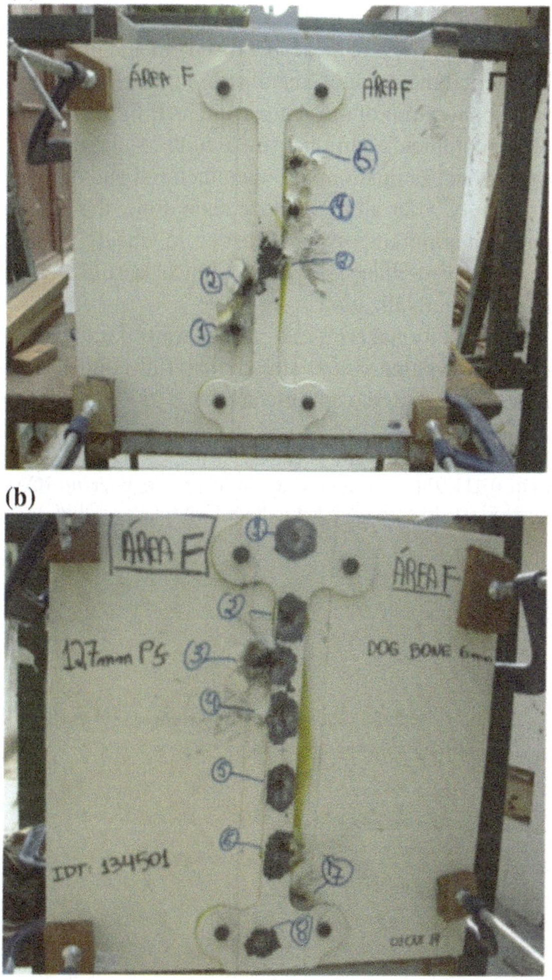

Fig. 4 Ballistic tests of the "dog-bones"

In the first stage one started the development of the manufacturing process, the definition of the various components, the preparation of the material specification sheets, the construction of the test specimens for physical and chemical analysis, mechanical tests and preliminary ballistic tests, development and elaboration of the different test panels for the various regions of the vehicle, in accordance with the protection required by the standards, as well as, in according to the list of tests prepared by the national authority, and finally, at this stage, aging tests were produced in order to simulating the working conditions of the vehicle for a long time of use. After this stage, the evaluation was started, carried out by the technical staff of the test center of the Army Evaluation Center (CAEx), on the representative test panels of the additional armor, as provided by Allied Engineering Publication, AEP-55, Volume 1 (Edition 1).

After approval of the tests carried out and its homologation by the national authority, started the stage of manufacture of the mock up of the panels that belong to the additional armor of the vehicle and its fitting adjustments. Subsequently, a prototype of the complete additional shield with its ceramic and steel panels for technical and operational evaluation was manufactured and delivered to the Brazilian Army.

Finally, after all these steps been done, the additional external armor kit for the medium armored personnel carrier on wheels, manufactured by ALLTEC, received from the Brazilian Army the technical certificates necessary for its manufacturing and commercialization.

The additional armored vehicle armor kit, developed and manufactured successfully, shows that Brazil already has full control of the technology to produce this strategic item in the defense area.

Acknowledgements We would like to thank to Financiadora de Estudos e Projetos (Finep) (Contrato 03.11.0341.00), Centro de Avaliação do Exército (CAEx), Diretoria de Fabricação (DF), Diretoria de Fiscalização de Produtos Controlados (DFPC).

Reference

1. STANAG AEP-55 (2005) Vol. 1—Procedures for evaluating the protection level of logistic and light armoured vehicles